AC6V's

Amateur Radio and DX Reference Guide

FM REPEATERS

FM101x

AC6V'S GUIDE TO VHF/UHF FM REPEATERS
AND
YOUR FIRST VHF/UHF RADIO

By Rodney R. Dinkins

AC6V Publications
981 Texas Rd
Iola, KS
66749

PREFACE

This book is intended for the new Amateur Radio Licensee who will be entering the world of VHF/UHF and FM Repeaters. If you are not yet licensed by the FCC, information on obtaining an Amateur Radio license can be found at URL: http://www.arrl.org/hamradio.html A Technician license is the first entry level, obtained by passing a 35-question multiple-choice examination. No Morse code exam is required for any class of license as of Feb 23, 2007. Many non-technical folks from all walks of life find the tests fairly easy and the license readily obtainable. This book does not prepare you for the Ham exams, but is intended for the new licensee getting on VHF/UHF repeaters.

Since entering the world of full time Dxing (retired) I have had the opportunity of listening to many of the FM repeaters in San Diego, Los Angeles, San Francisco, and on many trips around the country. When new hams come on the repeater, more often as not it requires several hours of tutoring by the old timers to learn about repeater operation, jargon, and protocol.

Listening for a long time will reveal some of the mysteries, but repeater jargon and technical terms often are still a mystery. After researching the available books on the subject, I thought the time is ripe for a new approach to the subject. This book is written for non-technical new comers to the repeater scene. No attempt is made to repeat the material on the ham exams, but rather the intent is to acquaint the new licensee with what goes on - on a repeater, terms, jargon, and how a repeater works from a user stand point.

Subjects include: Selection and Installation of your First Radio, Mobile and HT considerations, How Repeaters Work, How We Use Repeaters, Finding a Repeater, Offset, Split, CTCSS and PL, DTMF, DCS, Batteries and Power Supplies, Programming a Radio, Antennas, Repeater Myths, Jargon and Terminology, and much more. See the Table Of Contents. Since most readers will probably not read through the whole book continuously, subjects are sometimes repeated in several Chapters.

Rodney R. Dinkins AC6V
Oceanside, CA. E-Mail Address is on my web page http://ac6v.com/

CREDITS
Cover Antenna Photo – W6VR, Bob Gonsett and the Fallbrook Amateur Radio Society

My gratitude to the following contributors. KC6UQH, N6KI, WN6K, KG6DVD, W6ABE, N6FN, KM6K, W6YOO, W6VR, KG6JMP, W6EAJ, and my XYL Karla.

CONTENTS

CHAPTER 1: INTRODUCTION

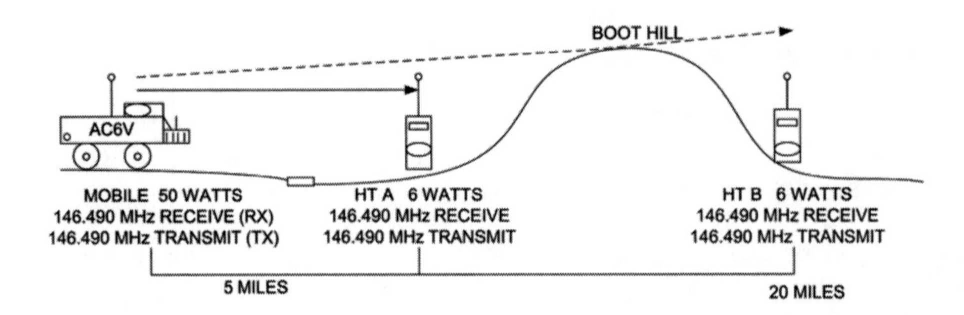

VHF/UHF radio is essentially line of sight. Above, the mobile wishes to communicate with Handi Talkies A and B directly without going thru a repeater. This is done using the same frequency to transmit and receive (simplex). Since the mobile and HT A are fairly close together with no obstructions they can carry out successful communications. But the mobile and HT B are too far apart and blocked by the hill and cannot hear each other. Using the illustration below, let's see how communication range can be extended using a repeater.

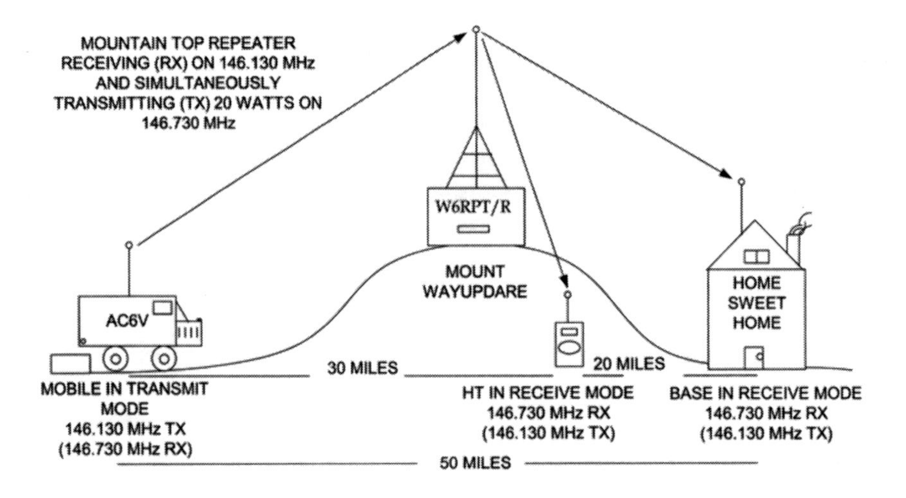

Here the mobile station wishes to communicate with both the base station and the pedestrian with a Handi Talkie. For direct communications, the three units are too far apart, and all are too low in altitude, as well as being blocked by the mountain. The repeater uses offset (split) receive and transmit frequencies and by virtue of its altitude, communications between all three units are achieved. How this all works will be explained in the following chapters.

As a beginner, you can jump on the repeater and learn the hard way with numerous folks explaining repeater usage (and arguing with one another as to what is proper). However, some VHF operators have very poor operating habits that you don't want to use.

Or you can read through all this and be a real pro when you enter the repeater scene -- your decision. Some folks are very helpful to beginners, but others give the old timer drill – "whatta ya doin on this repeater if you don't know what's goin on ??" Don't be put off by this, just a slice of life – this book will make it sound like you have been on repeaters for years. Since most readers will probably not read through the book continuously, subjects are repeated where necessary, so it serves as a reference guide.

It is not written for a particular locality other than the USA. See your local repeater web page or obtain handout sheets from your local repeater association for specifics of your local repeaters -- open, closed, autopatching, nets, protocol, commuters, usage, etc. Some repeater clubs have published guidelines, but many do not, I hope that this book will add to your enjoyment of repeaters and UHF/VHF.

Perhaps after reading the guidelines to protocol, it appears too formal or restrictive. This is not the intent, the whole purpose of this book is to acquaint the reader with what goes on - on a repeater -- terms, jargon, protocol, and how a repeater works from a user stand point. The guidelines contained herein are observations of repeater activities over a 25-year period. Perhaps once you know the drill, be as formal or as informal as you want within your repeater guidelines AND the FCC Rules -- see FCC Part 97 Rules, especially but not limited to: Identifying (§97.119), what constitutes third party traffic (§97.115), emergency communications (Subpart E), and business communications (§97.113). See URL:

http://www.arrl.org/FandES/field/regulations/news/part97

An appendix is included at the back of this manual to help you with the jargon and terminology. Perhaps Hams should just use plain language – but they don't. Conversations are full of jargon, terminology, and abbreviations. Most disciplines have their own jargon as does Ham Radio lingo.

One might hear "N4ZZZ this is K6XXX, Good morning OM, welcome to San Diego. Handle here is Jack, -- Juliet Alpha Charlie Kilo, QTH is El Cajon. You're not quite full quieting into the machine, about 20% path noise. Your deviation is fine. This repeater W6NWG Whiskey Six Nothing Works Good is located on Mount Palomar. The repeater gets very busy during commute hours so let's QSY to 146.075, plus offset with a PL of 107.2, QSL". This is followed by a beep, a quiet period, then the repeater drops off the air. Good Golly Miss Molly, what was that all about? We will cover all this jargon and terminology throughout the book, but here are quick answers so you don't have to flip through the Chapters.

K6XXX is the callsign of a California licensee; N4ZZZ was licensed in the USA 4th call district (southern USA). OM is an abbreviation for Old Man – jargon for "Fellow Ham". Handle is jargon for name. The Juliet Alpha Charlie Kilo are phonetics, similar to adam, mary, etc on the police frequencies. Hams use the world wide accepted NATO/ITU phonetics as a way to spell out letters that might be confused with others. QTH is a Q-signal, it means location. Whiskey Six Nothing Works Good are "funny phonetics" to make it memorable. A machine is a repeater. QSY means change frequency or channel.

Quieting is a characteristic of noise free FM. When the signal is strong enough – circuits clip off any static or noise, and this is termed full quieting. This is unlike AM or Sideband signals. A percent quieting is a subjective assessment of how much noise is on an FM signal if it is not full quieting. In frequency modulation, the audio you hear is proportional to the frequency swing or deviation, not amplitude as in AM signals. More on this later. In FM repeaters, your transmitting frequency is different than the receive frequency; the difference is called offset (or split) up or down (+ or -). More on this later. PL stands for "private line" which is a trademark of Motorola. It is more correctly called CTCSS. Some repeaters require a sub-audible continuous tone (PL) on your signal, without it the repeater will deny access. QSL is - Do you acknowledge (understand)? The beep is a courtesy tone – signaling the other station that is their turn to transmit. It also resets a repeater timer. More later. Well lots of protocol and jargon in that one short transmission. This book will elaborate on how repeaters and FM works and weed it all out for you.

CHAPTER 2: YOUR FIRST VHF/UHF RADIO

Typically the new licensee purchases a Handi-Talkie (HT) for VHF/UHF, but later find they are using their first radio mostly in a mobile or at a home base station. A mobile rig can be easily switched back and forth between the vehicle and a base location and offers considerable advantages over an HT. Keep in mind that the VHF/UHF radios are for the most part line of sight transmission and reception; however, repeaters can extend the range considerably. HF is the spectrum for long-range skip communications.

HTs are low power (usually 6 Watts or less) and most folks use battery power and thus constantly recharging batteries. Having two battery packs is almost a necessity. A battery pack can go dead very suddenly and you are off the air, not good during an emergency. If an HT is frequently used for base or mobile operation, consider using a DC power supply or an auto power adaptor. Be aware that the standard battery chargers supplied with the radio usually will not support transmitting, not enough capacity, and may result in hum on the transmitted signal.

Many HT users with a stock rubber duck antenna are forced to use maximum power to access repeaters and the batteries drain quickly. Because of the poor performance of a rubber duck antenna, depending on the users location, you can hear many HT users barely getting into the repeater and frequently dropping in and out. If you are close to repeaters, then an HT will generally work well. Also HTs operating at maximum power can get quite hot to the touch and curtail long transmissions.

Stock HT antennas although convenient are very inefficient, performance can be greatly improved with better antennas. However adapting an HT to mobile or base use can present problems. The typical HT is designed for rubber duck operation. When you put on a more efficient antenna, HTs can overload on strong signals and have severe intermod problems, especially in high RF environments like the big city. Intermod is short for intermodulation. This causes false or spurious signals that are produced by two or more strong signals mixing in a receiver or repeater station.

This is not to bad-mouth the HTs, they are great for carrying in your shirt pocket or belt clip for pedestrian use, public relations activities, and traveling light. Just be aware of their limitations and intended use. They certainly can be used for base and mobile applications, but better antennas and DC power sources are definitely something to consider when you settle in on an HT.

A mobile radio on the other hand can transmit up to 70 Watts, operates off of car DC power or a DC power supply at the base location and can use an antenna which typically is vastly superior to a rubber duck. Operating range will be considerably greater with the higher power and good antenna. Prices are about the same. So evaluate your needs and decide which starter radio is best for you. Many eventually end up with both mobile and HT rigs.

2 Meters is the most popular band and generally the least expensive first radio to buy. You may opt for a dual band 2M/440 transceiver if the 440 MHz band in your area has lots of repeaters. The 440 MHz band may give access to long range intertie systems and IRLP usage where these might not be available on 2 Meters in your area.

See Chapter 12 for the exciting new IRLP long-range communications systems. FM activity on 10M, 6M, 902 MHz and above are usually pretty quiet, depends on the area. 220 MHz can be active in some large metropolitan areas. Check your repeater guide and ask local VHF/UHF gurus for activity in your area.

Some go all out and buy one of the newer multi-banders which variously transceive on 6M, 2M, 220 MHz, 440 MHz, and for listening, cover the AM/FM broadcast bands as well as the 3-30 MHz HF spectrum. HT examples are the Yaesu VX-5, VX-7, Kenwood TH-F6 as well as others. You may find some of these rather difficult to program, so consider a programming cable and software.

If you have a general or extra class license or intend to upgrade, consider the new mobile/base multi-band, multi-mode radios that cover HF/VHF/UHF – with transmit in many bands and modes (CW, AM, FM, SSB, Data). Examples are the Kenwood TS-2000, ICOM IC-706 Series, and Yaesu FT-100D or FT-817 radios. Some of these have matching automatic antennas for mobiles. AC6V offers a book on HF and DXing, DX101x at URL: http://ac6v.com/DXSAMPLE.htm

Which brand to buy? See eHam reviews for topics on: Mobiles, HTs, Multiband and Multimodes, VHF/UHF Omni Antennas, and VHF/UHF Directional Antennas. URL: http://www.eham.net/reviews/ Talk to other local Hams for their advice, but beware of brand loyal types, best go see and try your choices up close and personal at the local radio store or borrow one from a friend for a trial.

A good way to decide upon your first radio is to make a checklist of the bands and features you expect to use. A checklist can be found at the end of this chapter. Downloading manufactures data sheets and researching the ARRL QST magazine reviews will prove helpful. A friend who is an ARRL member can download these reviews for you. See ARRL in the Glossary. Check the new equipment prices at the local Ham stores, major distributors, and web stores. If you decide to buy a used radio, you can minimize the risks by checking various sources for the current going price, E-Bay, swap pages, and used price guides, see URL: http://ac6v.com/swap.htm

For used radios, verify the condition of the radio and batteries, if it has the manual, and the original box and accessories. Obtain a written guarantee that the radio is fully functional in all modes and bands of operation. Money back guarantee if not. If sold as is – by no means pay the current going price, you may have some repair work ahead. Verify that the seller is a ham, and for mail order, that the address given to send funds matches the callbook address such as at www.QRZ.com. Be sure to evaluate the shipping costs, insured shipping, and which carrier to use. If so inclined, you may buy a broken radio and repair it yourself and save a lot of money. For example, some may be sold with bad batteries or a broken antenna.

Features to look for in VHF/UHF Radios

Power (Watts) Maximum power output typically ranges from 0.3 Watts to 6 Watts for HTs. The lower power units are OK for nearby repeaters. Mobiles – up to 70-Watt units are available.

Touch-Tone Pad – Allows you to use telephone interconnect (AutoPatch) and possibly access certain command functions.

Some radios have provision to memorize the autopatch codes and a long telephone number; this allows your favorite numbers and emergency numbers to be implemented at the touch of a single key or a few key presses. Tone stuff is covered in detail in Chapter 4.

Tone Encode (CTCSS OR PL) – Virtually a necessity for today's repeaters. An absolute must to access many repeaters, covered in Chapter 4.

Tone Decode (Tone Squelch) – Useful for blocking out unwanted transmissions. See Chapter 4.

RF Squelch - Allows user defined squelch level. See Chapter 4.

Memories – Can range from 30 to 200+ memories. The larger number being useful for those who travel.

Battery Types; NiCad - Nickel Cadmium, NiMH - Nickel Metal Hydride, Lithium, and Sealed Lead Acid are rechargeable types. Each has advantages and disadvantages. Also Alkaline batteries. See Chapter 6.

Alphanumeric readout – Allows for entering a name instead of the frequency. Example PARC for the Palomar Amateur Radio Club in place of 146.730 MHz. When programming lots of stuff in memory, the alphanumerics make it easier to recall the repeater name or area.

PC Programming – This allows for easy simplified setups of a transceiver by a computer program. Modern transceivers can be difficult to program. Computer programming is also very useful for travelers to set up a radio for a particular locale prior to an extended trip.

VOX – voice operated transmit. No need to press the Push-To-Talk switch (PTT) – just speak into the mic to transmit.

TOT - Time Out Timer. Turns off the radio at a predetermined time to avoid battery discharge. Also some TOTs will cut off your transmission if you talk beyond the TOT time. Usually the time is programmable.

REVerse – when listening to a repeater, you can press the REV button to hear the transmitting station direct (listen to the input frequency). You may hear them if they are close by.

Cross Band Repeat. Cross band repeating (CBR) is a feature with some VHF-UHF dual band radios that simply repeats what it receives on one band and automatically retransmit it on the other band. You can take an HT with you while hiking and transceive with your mobile radio which will amplify and cross band repeat to a repeater or other users. There are several methods of use, see Chapter 12.

Antenna Jacks – usually SO-239 for Mobiles and BNC on the larger HTs or SMA on the smaller HTs. For HTs, the BNC's are larger and much more friendly for external antenna connections. Some transceivers do not have provision for an external antenna. Adaptors are available to connect the smaller SMA connectors to BNC types.

Microphone Speaker Jacks. The ARRL suggests using a Mic/Speaker accessory as a safety feature. Hand-held radios are very popular for VHF and UHF operation, especially with FM repeaters. They transmit with less than 7-watts of power, which is generally considered safe. Because the radios are designed to be operated with an antenna that is within 20 centimeters of your body, they are classified as portable devices by the FCC. Some special considerations are in order to ensure safe operation. This is especially true because hand-held radios generally place the antenna close to your head.

Try to position the radio so the antenna is as far from your head (and especially your eyes) while transmitting. An external speaker microphone can be helpful.

Receiver Frequency Range – Many transceivers will receive well outside the Ham Bands making them useful for Military Affiliate Radio Service (MARS), Civil Air Patrol frequencies (CAP), or NOAA weather bulletins. Some cover the aviation bands; AM/FM broadcast bands, HF ham and shortwave frequencies.

MOBILE RADIOS - Same as HTs except car battery power is used and RF power outputs can be up to 70 Watts. These are adjustable to lower power values - typically, one, five, or 10 Watts. Mobiles typically offer better receiver performance. These can be used for a base station with a DC power supply. Detachable faceplate units are available for custom installation.

ALL MODE RADIOS can include AM, FM, CW, SSB, Data (packet and APRS). Prices are much higher. Be aware that hams (new or old) can get very frustrated with programming the "Do All" radios.

Ergonomics. Here we get into personal preferences. How does the HT fit your hand? Is the display easy to see (for you – maybe like me – old eyes)? Can the buttons be easily pressed? How about programming and button nomenclature – intuitive or gibberish?

Does it have a mic hanger so you don't sit on the mic and inadvertently key up a repeater? Is the antenna detachable, so you can put on an external antenna? Is the HT flimsy or built to drop and survive? "Micro talkies" can be very small and the displays hard to read – how you use these may decide what you buy. All this means that you should visit the local radio store or borrow one and try it.

Another way to evaluate a radio for ease of use and programming is to see the manuals on-line. For freebees - See URL: http://ac6v.com/manuals.htm For mobiles, ergonomics to be considered is the location of the microphone plug, right, left, or rear panel. Does the radio have a jack for an external speaker? Particularly if mounted in the dash. Radio Shack has external speakers that will give improved sound and volume. For open cockpit drivers, RS also has an amplified speaker. In some states, headphones are a no no when driving, some allow one earphone, so a single earpiece/boom mic can be adapted to your radio. Noise-canceling microphones can be used for cars, planes, boats, and motorcycles.

Most HTs tip over easily when set on a table or desk. A handy accessory you can build or buy for base use is a heavy base stand for the HT and use a combination speaker/microphone. See URL: http://www.niftyaccessories.com/ Another idea is to use a cell phone holder for your HT while in the car.

Mobile Radio installation. There is a temptation to mount a mobile radio in the dashboard and these give a nice professional look, but the radio can get very hot there and die. For in-dash installation or confined space, best get a "Mil-Spec" radio or one with a fan. Mil spec implies a radio designed to meet military specifications thus withstands harsher environments of vibration and heat. 12V outboard fans with a switch can be installed if you have the room in the dash. More expensive but more effective (from the standpoint of heat and adequate room) are the mobile radios with detachable faceplates, the radio proper can be mounted just about anywhere, behind the dash, in the trunk, or under a seat. Some cars have a transmission hump and accessories are available to mount the radio in a quick disconnect mount. One can snake coax under the dash or for roof mounts, route the cable through the windshield posts and underneath the headliner to the antenna connector.

Mobile antennas. Magnetic roof mounts are popular as no hole drilling or screw clamps are needed, but over a period of time paint scratches can occur. Trunk lid mounts or rain gutter mounts work well but here again paint damage or staining may happen over a period of time. Glass mounts have been improved to the point where they work well - but some autos have tinted passivated (metal particles) windshields which goof up the coupling -- ask your auto dealer. Glass mount antennas use capacitive coupling, so no hole drilling is required. But mounting your antenna near the engine can lead to ignition noise in the radio; however, suppressing ignition noise is usually an easy fix.

Roof Mounts. I like the mounts for roof installation, but these require careful installation. Drilling a hole in the roof can be dicey. Modern autos have thinner sheet metal than cars of yore and dimpling or detents may occur. Some use a hole saw, some a chassis punch, but professional installers have a tool especially designed for that purpose. Motorola makes one.

Be careful when drilling the hole – too much pressure can cause deforming of the roof contour. Keep in mind that you may have to deal with auto modifications when you sell the car. For roof mounts, rain caps are available or you can install a cell phone antenna for the new buyer. After any antenna installation, be sure and check the VSWR (Voltage Standing Wave Ratio) – see Chapter 4.

Noise. After installation, auto noise can interfere with the radio. There are two types of noise – received and transmitted. With FM radios, receiver noise such as ignition or computer noise are not usually a big problem, as it is in HF radio. Installing shielded wires can reduce ignition noise. Using laptop computers, which are well shielded, will reduce computer noise. Alternator whine is a transmitted noise and can be suppressed with a coaxial capacitor in the 12-Volt power leads. In the dire extreme, bonding (heavy strap wires or braid) of car components together may be required. Such as bonding the hood to the car frame or bonding the muffler to the frame, depends on the car.

For a thorough discussion of noise problems – see Automotive Noise Elimination By Stuart Downs. WY6EE URL: http://www.arrl.org/tis/info/pdf/001qex32.pdf

Mobile power. It is tempting to use the cigarette lighter plug for 12 Volts to power the mobile radio and this may work for low power units, but it is always best to run two heavy gage wires through the firewall and directly to the battery via fuses in both lines. Don't use the chassis as a ground return, ground loop problems and noise pickup may occur, run both + and – lines to the battery. FM mobile transceivers with 50 W of output power can draw up to 10-12 Amperes of current at 12 Volts. Most cigarette lighter plugs cannot handle these high currents. Operating a 50 W mobile transceiver from the cigarette lighter plug may cause permanent damage to vehicle's electrical system, and can also be a fire hazard. You can probably operate a 6 W handheld transceiver from the cigarette lighter plug, but not a higher wattage radio. Run the power wiring as far as physically possible from the ignition system, the engine-management electronics, and of course, the engine. Use plastic cable ties to securely fix the wiring to already fixed fittings in the engine bay.

If you have considerably dimming of radio lights and the display when transmitting at high power, it is probably caused by too much IR drop across the power leads, be sure to use the heaviest gage wire practicable. Also allow for some flexing due to vehicle movement. Dimming could also be caused by a bad electrical connection somewhere between the vehicle battery (or a power supply) and the transceiver. Always use solder connections and solder lugs where required. Use heat shrink tubing not friction tape, as tape tends to unravel with heat. For quick disconnects, a variety of connectors can be used.

This will allow you to take the radio in the house for base use or testing. For a great pictorial step-by-step mobile installations – see URL: http://ac6v.com/opmodes.htm#MO

Finally after installing a mobile rig, be sure it does not interfere with the auto electronics. Cases have occurred where RF has interfered with brake system electronics and other devices. RF bypassing will be in order and many auto manufactures have kits for these problems.

COAX. The transmission line between the radio and antenna has several considerations. If you have more than 3 dB of loss in the transmission line, best install the next level of low loss cable.

Coax loss charts are available on the web – see URL: http://www.k1ttt.net/technote/coaxloss.html#tables or you can use a power meter at the radio, and then measure the power at the antenna. Remember 3 dB loss is half power. A rough guideline is -- at 144 MHz (2M), with type RG-58, about a 25 foot run is the limit. With RG-8 it is about 100 feet. At 440 MHz (70 cm) the lengths are about half.

For outside base antennas, be aware that air dielectric cable can collect water. Also, many coaxial cables cannot support their own weight for long runs. So coax ran up a tower or mast is usually secured at 3 or 4-foot intervals, less with foam dielectric types. Coax cables on high towers have been known to break down under the stress and short.

Also be aware that coax cable has a maximum recommended power handling capacity. So if you are running power above 100 watts – best check the charts to be on the safe side. If you have some amount of VSWR, the power handling capacity of the coax should be de-rated. Power handling capacity is inversely proportional to SWR. The life of a coaxial cable depends on many factors. Some of those are ultra-violet exposure, migration, high humidity, moisture, age, corrosion, power/heat, and voltage. Coaxial cable can perform to its maximum designed efficiency about an average of seven to ten years.

If your signal or VSWR is erratic or you are unable to get the maximum performance from your transceiver, then it may be time to consider changing your coaxial cable or cable assemblies. Putting a dc ohmmeter on coax will tell if the dc paths are OK, but tells nothing about losses when RF and power are sent down the cable. One good way is to send max power down the coax and measure the power at the transmitter, and then measure it at the antenna. The cable loss should be in spec.

CHECKLIST (See Glossary if in doubt about the feature).

Budget $ _______ New ___ Used ___ Broke and you repair ____

Transceive Bands & Power (Watts) 2M __ 440 MHz __ 220 MHz __ 6M __ 10M __ 1.2 GHz __ HF __

Bands: Monoband ___, Dualband ___, Multiband VHF/UHF ___, Multiband HF/VHF/UHF ___

Modes: FM __ AM __ CW __ SSB __Data __ Crossband Repeat ___ VOX ___ Memories (Number) __

Receiver Frequency Range: MARS__ CAP__ NOAA weather __ Aviation bands __ AM/FM __ SW __

Touch-Tone Pad___ Can store long telephone numbers __

Tone Encode/Decode (CTCSS OR PL) __ DCS __

Features: Alphanumeric Readout __ REVerse Button __ Open Squelch Button __

Mobile __or HT __ Micro Talky or Big HT ____ Is the HT flimsy or built to drop and survive ____

Mobile Antenna: Roof Mount __ Trunk Lid __ Rain Gutter __ Mag Mount __ Glass Mount ___ Other _

Mobile Radio: Detachable Faceplate __ Coax Type ____ External Speaker ___ Amplified Speaker ___

HT External Antenna: Improved Duck __ Collapsible Antenna ¼ ___½ ___or 5/8 wave ____

HT Accessories: – auto cigarette lighter adaptable ____ HT stand for base use ___ Time Out Timer ___

Antenna Jacks: SO-239 ___ BNC ___ SMA ___ None ___ Adapters for BNC to SMA or vice versa __

Accessories: Quick Charger __ Auto charger __ Speaker Mic __ PC Programming Cable ___ Other ___

Accessories: Boom Mic __ Earphones(s) __ Noise Canceling Mic ___ Mic Wind Screen ____ VSWR Meter __

Battery Type _________ Batteries Supplied ___ Battery Shell Adapter for Alkalines ___

Ergonomics: How does the HT fit your hand _____ Can the buttons be easily pressed ____

Ergonomics: Is the display easy to see _____ Are Face Plate & Buttons & Mic Buttons Backlit ____

Programming and button nomenclature: intuitive ___or gibberish _____

Speaker & Mics: Speaker output jack ____ Mic/Speaker Jacks ____ Mic connector location ____

Reliability: Mil Spec radio __ Fan __ External Fan _____

VOIP such as WIRES II, D-STAR – See Chapter 12

Cheat Sheet _______

Other Considerations ______________________________________

NOTES

CHAPTER 3: OPERATING SIMPLEX

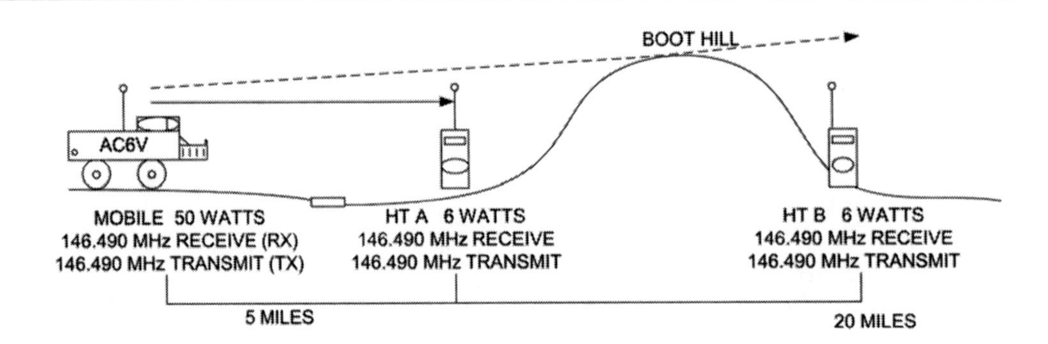

The mobile wishes to communicate with Handi Talkies A and B using the same frequency to transmit and receive (simplex). Since the mobile and HT A are fairly close together with no obstructions, they can carry out successful communications. But the mobile and HT B can not hear each other, as they are too far apart and blocked by the hill.

If you are in line of sight to another station with no obstructions, you may be able to operate direct on the same frequency (simplex) to another station (or stations). For example, while on a repeater, you and the other station are close by and the repeater is very busy, so you can switch to a simplex channel. For example, 146.490 MHz -- both stations set their VFO's (or bring it up out of memory) for that frequency. Here you will alternately transmit and receive on the same frequency – no offset. Operation is half duplex, that is one station transmits and the other listens, only one can talk at a time. You can operate in FM, SSB, and several other modes if your radio has these capabilities. If simplex doesn't work – try another repeater. Many clubs have more than one repeater, one of these is usually a "prime" repeater, and the others may be relatively quiet.

Most rigs have two tuning modes, VFO and Memory. VFO stands for Variable Frequency Oscillator – about the same as the tuning dial on an AM or FM broadcast radio. VFOs in modern radios are digital (as opposed to analog – like the old AM radio slide rule dial). The VFO mode allows you to dial in frequencies in the radio's range. For digital VFOs, this is not continuous tuning but in steps, usually provision is made to tune in large or small steps. Step increments are a selectable menu function, but the radio factory default setting is fine for most usage. In the memory mode – the frequency and associated data are programmed in and cannot be changed without reprogramming. One merely calls up a channel and you are set to go. Once programmed into memory, no need to fiddle with frequency, offset, etc., they are all stored in a memorized channel. Up to 200+ memory channels are typical.

For a simplex channel, first find the manual page to access the VFO and put in the desired frequency. This allows you to manually tune a frequency into the HTs Variable Frequency Oscillator. Write down how to get in this mode. Next, find the page for offset – write down it down. Check that the offset is neither plus or minus and is in the simplex setting, and then memory write all. See Programming – Chapter 5. You can invoke PL encode and decode or DCS if you prefer – see Chapter 4.

When first setting up a radio, in addition to your favorite repeaters, program in a couple of simplex channels including the National simplex calling frequency. Calling frequencies are gentleperson agreements and ARRL recommendations for the favored simplex channels for folks to call on VHF/UHF when repeaters are out of range.

Once contact has been made on a National calling frequency, it is best to switch to another simplex channel or repeater. This clears the National calling frequency for its intended use. So always have several simplex channels in memory. Base stations can easily dial in simplex channels, but a mobile may be too occupied with driving to fiddle with setting in a new simplex frequency. Simplex favorites for 2 meters are 146.490, 146.550 and 147.555 (referred to as "triple nickels"). Varies with area.

For simplex channels it is best for both parties to ask if the frequency is in use or has another use such as a remote base. You may be in the clear but the station that you are contacting may interfere with a station up the line that you can't hear. Just give your callsign and ask, "Is the frequency in Use?"

The display on some rigs show an "S" for simplex, others show an absence of a + or - offset sign, still others have an RP+ or RP- sign. Because simplex usually has a limited range it can give a degree of privacy but not a lot, some one always seems to be listening, so be careful of gossip. You may be surprised at the hi-fi sound of simplex as most repeaters cut off a certain amount of the high and low frequencies. Typical audio range of repeaters is 300 to 3,000 Hz. Usually wider when operating simplex.

The communication range between Amateur VHF/UHF FM mobile and hand held radios at ground level, operating simplex (direct) is about five to fifteen miles for mobiles, and just a couple of miles for hand held transceivers. The range depends on the band of operation, transmitter powers, antenna heights, obstructions, antenna gains and receiver sensitivity or noise figures. Essentially it is line of sight. See Guide to VHF/UHF Transmitter Range By Artsci. http://www.artscipub.com/simpleton/simp.range.html

Naturally, trees, buildings, hills, gullies, mountains, and obstructions can attenuate or block signals. In addition, reflections from buildings, mountains, overhead aircraft, tunnels, etc. can give some weird results such as multi-path reception or distortion. Remember flutter caused by airplanes on TV ?

However, skilled VHF operators routinely make contacts that exceed the geometric horizon. This usually requires directional antennas high atop the house or a tower and using higher power levels. Several techniques are successful, knife-edging, peak bouncing, ground reflections and others. Knife edging involves the signals contacting a hilltop and bending downwards for increased range. Especially when using modes such as CW or SSB.

Another is peak bouncing where a directional antenna is pointed elsewhere from the desired direction and bounces off a peak and gets to the desired direction. Local ducting can occur which can extend the range considerably. But ducting can come and go – so not a reliable communication path. Certain tropospheric propagation modes can occur and the range can be phenomenal, hams in California have worked into Hawaii on 2 meters and higher frequencies – but this is very rare and not the norm.

Some tips on operating simplex include using better antennas than a rubber duck and increasing power until adequate communication is established. A vertical antenna "talking" to a horizontal antenna will exhibit cross-polarization which results in significant attenuation, so endeavor to have the two stations use the same type of polarized antenna.

OPERATING ON VHF/UHF SSB AND CW

The all mode radios will allow you to operate on Single Sideband (SSB), AM, and CW in simplex on all the VHF/UHF bands. Except for contests, activity can be very low, but may be a good place to have a semi-private QSO. Typically horizontal antennas are used, but you can use verticals as well. Just be aware of cross polarization attenuation when operating line of sight. SSB convention is upper sideband (USB). Protocol and usage is very similar to the HF bands. VHF/UHF contests are conducted periodically and this is a good time to make contacts. See Contests URL: http://www2.arrl.org/contests/calendar.html
For operating frequencies, see Calling Frequencies http://ac6v.com/callfreq.htm

6 Meters is called the magic band and one can experience exciting F layer skip and contacts can be made all over the USA and to other countries. However F layer skip occurs only during the peak of the 11-year sunspot cycle and only for very high values of the solar flux index. Sporadic E skip can occur at any time and is more localized, providing contacts to near by states The band can be essentially line of sight and very quiet for long periods of time. Prime time for F layer skip if it occurs is usually the winter months. Sporadic E is more likely in the summer time. Don't expect too much of a 6M HT with a rubber duck for working skip, a much better external antenna will be required. With the advent of the new multi-band radios, more repeater activity is taking place on 6 Meters.

The best way to work 6M DX is to put the DX windows in memory - 50.06-50.09 Beacons, 50.090 CW Calling Frequency, and 50.110 SSB DX Calling Frequency. Put these in memory -- then scan them with the squelch just engaged. Or scan the entire SSB DX window - 50.100 to 50.130. Another way is to monitor the DX packet cluster – 6M enthusiasts will scan the band periodically and post any domestic or DX activity. If the band is open, it will appear on the clusters. DX Packet Clusters are now available on the web via telnet, see URL: http://ac6v.com/dxcluster.htm#TN Also many areas have VHF DX Packet Clusters- check with your local DX Club.

50 MHz Frequency Usage and Calling Frequencies See URL: http://ac6v.com/callfreq.htm
50.100 to 50.130 DX Window (USB)
50.110 DX Calling Frequency (USB)
50.125 National SSB Simplex Frequency (USB)
50.4 National AM Simplex Frequency

50 MHz (6 METERS) FM SIMPLEX VOICE FREQUENCIES

51.500	51.520	51.540	51.560	51.580	51.600
52.490	52.510	52.525*	52.550	52.570	52.590

* 52.525 is the National FM Voice Simplex Calling Frequency

---

144 MHz (2 METERS) FM SIMPLEX VOICE FREQUENCIES

146.400	146.415	146.430	146.445	146.460	146.475
146.490	146.505	146.520*	146.535	146.550	146.565
146.580	146.595	147.405	147.420	147.435	147.450
147.465	147.480	147.495	147.510	147.525	147.540
147.555	147.570	147.585			

*146.520 is the National FM Voice Simplex Calling Frequency

144 MHz (2 METERS) Digital/Packet Simplex Frequencies - 25 Simplex Channels

144.910	144.930	144.950	144.970	144.990	145.010	145.030
145.050	145.070	145.090	145.510	145.530	145.550	145.570
145.590	145.610	145.630	145.650	145.670	145.690	145.710
145.730	145.750	145.770	145.790			

220 MHz (1.25 METERS) FM Voice Simplex Frequencies

223.400	223.420	223.440	223.460	223.480	223.500 *	
223.520	223.540	223.560	223.580	223.600	223.620	223.640

* 223.500 is the National FM Voice Simplex Calling Frequency

440 MHz (70 CENTIMETERS) FM SIMPLEX VOICE FREQUENCIES

445.9125	445.9250	445.9375	445.9500	445.9625
445.9750	445.9875	446.0000 *	446.0125	446.0250
446.0375	446.0500	446.0625	446.0750	446.0875
446.1000	446.1125	446.1250	446.1375	446.1500
446.1625	446.1750			

* 446.0000 is the National Simplex Calling Frequency

906.000 to 907.000 MHz (33 CENTIMETERS) FM SIMPLEX VOICE FREQUENCIES

906.000 to 907.000 MHz - channel every 25 kHz
906.500 is the National FM Voice Simplex Calling Frequency

1.2 GHz (23 CENTIMETERS) FM SIMPLEX VOICE FREQUENCIES

1294.00 - 1295.00 Narrow Band FM simplex, every 25 kHz
1294.50 is the National FM Voice Simplex Calling Frequency

CHAPTER 4: HOW REPEATER OPERATIONS WORK

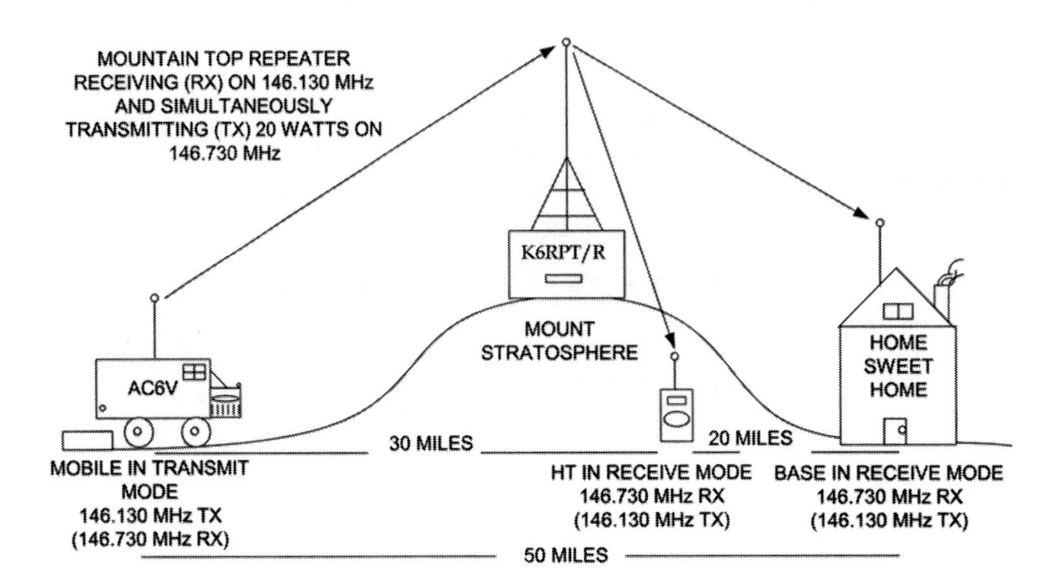

Repeaters work using offset frequencies and lots of tone stuff. Following are the details on the operational aspects of repeaters. In Chapter 7 we will discuss how we use repeaters for conversations, emergency use, and nets. As you encounter unfamiliar terms, refer to the glossary included at the end of this book.

In the illustration above, the mobile station wishes to communicate with both the base and the pedestrian with a Handi-Talkie (HT). But the mobile, base, and HT are too far apart, too low in altitude and blocked by the mountain. So the mobile transmits on 146.130 MHz to the mountain top repeater. The repeater processes the FM signal and simultaneously transmits the information on 146.730 MHz. Both the base and HT are in the receive mode (146.730 MHz) and pick up the communication from the repeater that was transmitted by the mobile station – slick huh?

When the HT wishes to transmit (when the mobile station is done), the user presses the mic button and the HT will automatically switch to transmit mode and transmit at 146.130 MHz, the other two stations, when not transmitting, automatically go to the receive mode and receive the communication on 146.730 MHz. You'll may hear this repeater referred to as "The 73 Repeater or the Mount Stratosphere machine".

VHF/UHF is essentially line of sight. The three stations might have a direct range to each other of a few miles line of sight (5 to 10 miles), but can send a signal many more miles in the up direction to the mountaintop. Repeater ranges of 100 miles plus are not uncommon depending on the altitude (and other factors) of the repeater and the other stations. As an example, a repeater here in San Diego is located on a mountain at 5600 feet and covers about 2,000 square miles of San Diego county and 100+ miles into the Pacific.

Note that even though the respective stations are transmitting power levels in the Watts range, the signals on the receive side are very small, in the microvolt range. The repeater and your radio do the work of processing and amplifying the signals, perhaps the repeater doing most of it, by virtue of its altitude. If you are close to a repeater, a few tenths of a Watt may be sufficient to access a repeater, even with a handheld and a rubber duck antenna. Further out, more power and/or better antennas will be required.

Be aware of repeater characteristics. Some are directional, maybe covering North, East, and West, but not down south, so as avoid interference to other areas. Some are "elephants" – a repeater that receives further than it can transmit, big ears, and small mouth! Or an "alligator" – a repeater that transmits further than it can receive, big mouth, small ears! The alligator concept will answer the question " Jeepers, I have the xyz repeater very strong - over S9, why can't I access it even with the proper PL and 50 Watts?"

Repeater With Small Ears - Big Mouth . You can hear it but can't access it !!

Repeater With Big Ears And A Small Mouth. You can access it but can't hear it !!! How would you know this? Some one will tell you for sure.

Some repeaters (machines) are high level (mountain top) covering well out in range; others are low level (building top) covering only a local area. Ask on your local repeater about the area repeaters, coverage, frequency and location. There are always repeater gurus who know of every machine in an area and its characteristics. Later we will discuss how to find repeaters in your area or while traveling. If you want to know more about transmitter range – see URL: http://www.artscipub.com/simpleton/simp.range.html Also a line of sight calculator can be found at URL; http://ac6v.com/repeaters.htm#TR

Repeaters are owned by individual amateurs or more commonly by repeater clubs. All costs and maintenance are incurred by the repeater owner or by dues from club members. Some repeaters are closed to members only. Many are open to any amateur. For frequent users, membership is usually encouraged. If you use a repeater frequently, it is a good idea to become a member of the repeater club; clubs need help in defraying the costs of operation.

Repeaters can have many features beyond just extending the range of mobile or handheld radios. Digital Voice Mailboxes can be implemented and have the ability to record and store voice messages for users. Some have synthesized speech for ID, time, and weather. Some repeaters have a Bulletin Board feature where club meeting announcements and other info is available by using touch-tone commands. One especially useful feature is called Autopatch. This is a telephone line and special control equipment at the repeater that allows you to make local phone calls from your radio. Not a cell phone by any means, as you can't do business on Amateur Radio, including on the autopatch. See Chapter 7, Using Repeaters

You can't receive calls, you can make only local calls, and your conversation is not private! Everybody listening to the repeater hears your conversation. Still, autopatch is handy, within its limitations. And the conversation is "one way (half duplex)" you can't hear your party while you are talking.

Be sure to read the FCC rules involving autopatching, see FCC Part 97 Rules, as to what constitutes third party traffic (§97.115 and business communications (§97.113).

SIMPLEX, HALF DUPLEX, FULL DUPLEX

Simplex - a communications mode in which radios transmit and receive on the same frequency. Only one station transmits while the other station receives and vice versa. FRS and CB radios are an example of simplex radios. There are such things as simplex repeaters, but these are rare and not discussed here.

Half duplex (Semi Duplex) - a communications mode in which a radio transmits and receives on two different frequencies but performs only one of these operations at any given time. In half duplex, only one station can talk at a time. Your VHF or UHF radio is operating half duplex when set up for standard repeater use. Example you transmit on 146.13 MHz (can't hear the other station while transmitting). Then you listen on 146.73 as the other station transmits. Your radio automatically transmits on 146.13 MHz when you press the PTT switch and reverts back to 146.73 MHz to listen when you release the PTT.

Full duplex - A telephone is a full-duplex device because both parties can talk at the same time. For radios this is a communications mode in which a radio can transmit and receive at the same time by using two different frequencies, usually on different bands. Some VHF/UHF radios can transmit on VHF and simultaneously receive on UHF for satellite work. And some sophisticated multi-repeater systems and inter-tie systems are capable of full duplex

REVerse – when listening to a repeater, you can press the REV button to hear the transmitting station direct (listen to the input frequency). You may hear them if they are close by. Mobiles will typically cause your S-meter to bobble around.

OFFSET, SPLIT, INPUT AND OUTPUT FREQUENCIES

Repeaters are not parrots (listen then repeat), so repeat is perhaps a misnomer, as it doesn't listen, store, and then after a delay re-transmit, it does this simultaneously. For repeater operation, our transceivers have to be preset or programmed to transmit on one frequency and receive on a different frequency. This is referred to as offset, split, input/output. See conventions for Offsets at the end of this chapter.

For an offset example, consider a repeater that hears you as you transmit on 146.130 MHz (input), and repeats you onto 146.730 MHz (output). The amount that the receive frequency is offset from the transmit frequency is called just that -- the "offset" or "offset frequency". Sometimes referred to as "split". In this case a 600 kHz offset. Since this repeater receives below the transmit frequency it is termed a minus offset. A repeater that receives on 147.130 and transmits on 147.730 MHz has an offset of plus 600 kHz.

Repeaters that have outputs in the lower part of the 146 MHz portion are often plus offsets while those operating in the upper portion of 146 MHz are usually minus offsets. For the 145 MHz range – splits are minus, and for the 147 MHz range – splits are plus. See your repeater guides. Most of the new rigs default to the standard offsets. In some areas, repeaters operate on a reverse split – check you repeater guide. Although 600 kHz is a standard offset for 2 Meters, other "oddball splits" are sometimes used, but rarely.

An easy way to remember offsets is (+) Plus Offset = Transmit UP. (-) Minus Offset = Transmit DOWN

For offsets conventions on the other bands, 10M. 6M, 440 MHz, 220 MHz etc., see offset conventions at the end of this chapter.

VFO, MEMORY, CTCSS, PL, TONE SQUELCH, DCS, DTCS, DTMF, TONE BURST, RF Squelch

Most rigs have two tuning modes, VFO and Memory. VFO stands for Variable Frequency Oscillator – about the same as the tuning dial on an AM or FM broadcast radio. VFOs in modern radios are digital (as opposed to analog – like the old AM radio slide rule dial). The VFO mode allows you to dial in frequencies in the radio's range. For digital VFOs, this is not continuous tuning but in steps, usually provision is made to tune in large or small steps. Step increments are a selectable menu function, but the radio factory default setting is fine for most usage. In the memory mode – the frequency and associated data are programmed in and cannot be changed without reprogramming. One merely calls up a channel and you are set to go. Once programmed into memory, no need to fiddle with frequency, offset, etc., they are all stored in a memorized channel. Up to 200+ memory channels are typical.

CTCSS Continuously Tone Coded Squelch System, also known as Subaudible Tone and "PL" (Private-Line, a Motorola trade name). Commonly used for repeater access. These are specific frequencies between 67 and 254.1 Hz. Hereafter referred to as PL for shortness and common use. PL has encode and decode functions. Encode sends PL to the repeater. Decode is set in your receiver to detect the PL from the repeater - see Tone Squelch below. PL frequencies and corresponding ICOM and Motorola numbers are given at the end of this chapter. In Amateur Radio, many repeaters require users to send the correct PL tone continuously to use the repeater. In fact, some coordinating groups insist on Pled repeaters. This may mean the repeater is "closed," for use only by members, or it may simply be used to avoid being keyed up by users of another repeater on the same frequency pair. Usually PL is required for phone patching.

Some of the new radios have an automatic PL seeker, typically referred to as CTCSS Tone Search or a similar term -- check your manual. Your brand of radio may or may not have all standard PL tones. CAUTION: Some rigs have a PL factory default (e.g., 88.5 Hz), be sure to change this when accessing a non Pled repeater, you could inadvertently bring up another repeater down the line that is Pled at 88.5 Hz.

Except for riding the squelch tail briefly, you CAN NOT get into a repeater that is using PL access -- UNLESS your radio is set for the correct PL tone. Check the repeater guides to see which repeaters require PL. When setting up your radio, be sure to do three things, 1) set the PL frequency, 2) activate the tone -- a little "T" or "Tone" indicator should appear on the front panel readout and then 3) memory write all. See Chapter 5 – Programming. Although very rare, some repeaters use non-standard PL's for their own reasons (privacy). If you can't access a repeater, a quick troubleshooting list is given in Chapter 7.

TONE SQUELCH. Also referred to as Tone Decode. A receiver function -- an adjunct to PL to silence your receiver if the proper PL is not received from the transmitting station. Some repeaters transmit PL all the time whether they receive PL or not. This allows some radios to invoke Tone Squelch, (T.SQ). With this, your receiver listens for the proper repeater's PL tone, and if not detected, silences your receiver. Operation is independent of whether the repeater requires PL for access or not. This is useful where two or more repeaters are operating on the same frequency with different PL tones. On repeaters it is usually the same frequency as for accessing the repeater

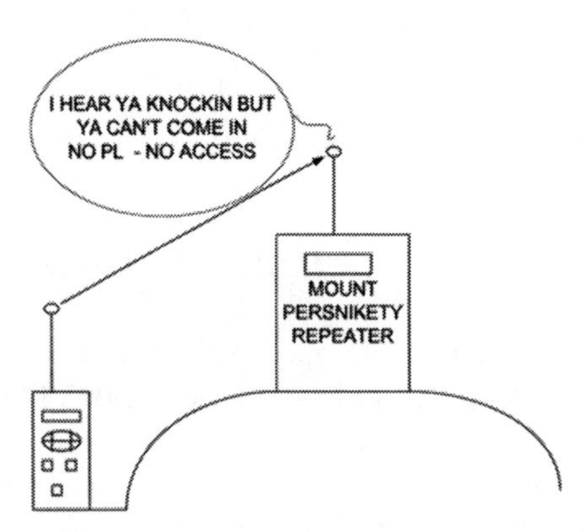

DTMF Dual-Tone Multi-Frequency. (Or Touch-tone an AT&T trade name) similar to telephone touch-tone pads.

DTMF is used all over Amateur Radio, for autopatching phone calls and for remote control of the repeater by control operators. Autopatch codes are usually available only to repeater members -- check with the club trustee for these codes, rules, and procedures.

Always identify with your callsign before sending any DTMF codes. Some radios have a paging function with DTMF to call a particular station or group of stations. Appears this is similar to Alinco DSQ. Also some radios have provision to memorize the autopatch codes and the long telephone number, this allows your favorite numbers and emergency numbers to be implemented at the touch of a single key or a few key presses -- again see your manual.

DCS (DTCS) [Digital (Tone) Coded Squelch]. A somewhat "private" calling system which can be used on some repeaters, but it is more commonly used on simplex for busy channels (e.g., Dayton Ham Fest). Appears Kenwood and Yaesu calls it DCS and ICOM's term is DTCS. Also DCS may have other variations such as Yaesu ARTS. It is digital data or code words that are transmitted with or without the voice audio at frequencies below 300 Hz. Not all Ham rigs have these features - check your manual. Nor do all repeaters support this function – ask the repeater guru. DCS will squelch down a receiver until the proper codes are received thus is used to silently monitor busy channels. Not the same as Tone Squelch.

Unlike CTCSS, which uses a continuous tone, DCS uses code words, which are unique, and all code words may be used on the same channel without interference. There are 104 standard DCS codes -- see your manual. Your brand of radio may or may not have all of the standard DCS codes. To avoid transmitting on top of another transmission, one must take care to observe the S-Meter and/or the "Busy Light". Both Ringer and Tone Scanning functions are available with some systems.

Tone Burst Some repeaters and radios use a tone burst system which is typically a 1,750-Hertz tone burst to access repeaters. This is an older system not much used in the USA anymore. Unfortunately, some radios default to this and send an annoying tone burst even though your repeater doesn't require it -- check the manual to turn it off if it is not required. Repeater cops will bug you if the tone burst is on.

RF Squelch. Keeps the receiver squelched until the incoming signal exceeds a user-defined signal strength (S-Unit). The RF squelch gives a qualitative setting. Normal audio squelch is set "by ear". When weak signals cause the squelch to open frequently, if you don't want to use CTCSS or DCS control, the RF Squelch feature will keep your receiver quiet until a stronger signal is received. Handy when scanning the band.

TIMERS, BEEPS and BE-BOPS.

Beeps. Repeaters use a courtesy tone or beep to signal when a station has quit transmitting and it is the next station's turn. Also at the beep the timer is reset. Following the courtesy tone, a brief delay is heard before the repeater drops out. Below is a TYPICAL repeater-timing diagram, however PLEASE NOTE that various repeaters have variations on this and the times. Your transceiver may also have beeps or be-bops to signal key presses, band edges or power on-off.

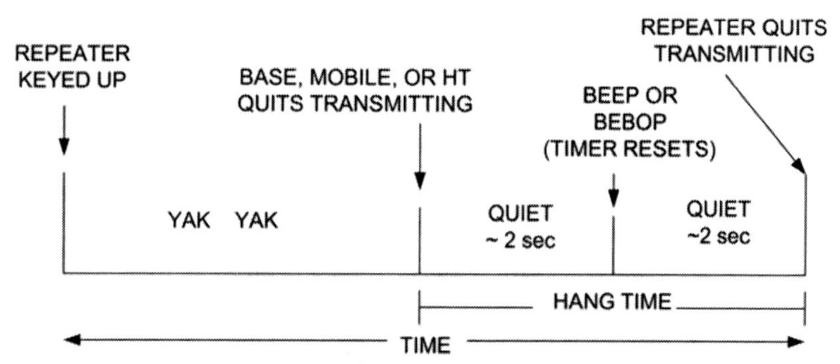

Be-Bops Some systems use a two tone or be-bop tone as a courtesy tone and may also signal various conditions of the repeater control functions (e.g., controller in unlocked status). Inter-tie systems may use be-bop as a courtesy tone and may use it as an indication of the system status. At times special beep sequences such as multiple beeps (beep beep beep) are heard -- for meanings, see your local repeater guidelines. Examples, one repeater here has three operating modes, which are signified audibly by courtesy tones generated after the transmitting station releases the push-to-talk switch. The modes are as follows: Normal - Single Beep; Club Net - Double Beep; ARES Net - Triple Beep. Obviously one should refrain from using the repeater if in the Club Net or ARES Net mode. On the other hand, another local repeater here uses a triple beep to signal a repeater timeout. And some repeaters use a "bubble-up" tone for the courtesy tone. Some have a special tone or tones sequences to indicate that the autopatch is accessible or shutdown.

Roger Beep (R) (or Morse K) -- Over or Go Ahead Beep - Occasionally a station will come on with a Morse dit-dah-dit (R) or dah di dah (K) roger beep or over to you beep. This is usually from an after-market microphone and smacks of CB operators - not a good idea.

TIMERS. Almost all repeaters have timers. A timer (sometimes called a QSO timer) is a clock that starts counting when you begin to transmit through the repeater. Typically, this clock is set to "time-out" after about a minute to three minutes. That means that if you transmit continuously through the repeater for more than the timer time, the repeater will go off the air (we call it "timing out the repeater" or the "alligator got ya"). Your transceiver may have a timer also to turn off the radio after an idle period.

Repeater timers usually reset to zero when you hear the courtesy tone or beep. So you must keep your transmissions under the repeater timeout period, and always wait for the beep, to avoid having your transmission dumped by the repeater timer. Also some transceivers have a TOT (Time Out Timer), which can be adjusted from three to ten minutes to "self" cut off your transceiver so you can't talk too long.

Note that transmission times are accumulative if folks don't wait for the beep. Example: A machine has a 60 second timer and station X talks for 50 seconds, if station Y comes in before the beep, 10 seconds later the timer will shut down the repeater. To signal a timer dropout, some repeaters use a voice announcement, others use a series of short beeps, such as a triple beep. Others may use a short delay, a short beep and then dropout.

Typically after timeout, there may be a several second wait before the repeater can be brought back up on the air. If the offending talker or a carrier persists, the repeater will remain mute until the input signal disappears, so timeouts can be quite long. Common sense will take over after you have a little repeater operation under the belt, and you won't even have to think about when to ID or when to drop your transmission - you'll operate automatically.

On some repeaters, you will hear a squelch tail – which is a burst of noise before the repeater drops out. Many modern repeaters have an anti-squelch circuit so this isn't heard.

WE PAUSE FOR STATION IDENTIFICATION...

Repeaters must identify every 10 minutes to satisfy FCC rules. Most repeaters identify in Morse code (but not over 20 words per minute) and some use a voice synthesizer. Those that use a voice ID, may also send a greeting, PL frequency, time, temperature, and notification of events, such as the next club meeting. The format for IDing is determined by the microprocessor code in the controller and can be set in a variety of ways and varies from repeater to repeater. Some send the Morse ID twice, one at 5 WPM, then 20 WPM. Some repeaters will send Morse K after the callsign, (Dah Dit Dah) this means invitation to transmit.

On various repeaters, there can be both short ID's and long ID's (or just one ID). A short ID is the repeater call sign followed by repeater (or slash R in Morse) and comes up at 10-minute intervals or less when the repeater is in use. There can be a "long ID" which sends the ID and typically the location of the repeater. This typically comes on if the repeater has not been used during a 10-minute period or the period of the ID timer. This will vary, as repeaters can be set up just about any way the repeater owners desire, within the FCC rules. Example, a local repeater sends the long ID only every fourth ID cycle.

When a repeater is IDing, best not to talk over the repeater IDer as some Ham might be listening to determine the repeater callsign and location

The RULES say YOU must ID once every 10 minutes and when you sign off. Also best to give your callsign when you first come on so folks know who you are. No need to overdo it however, some folks give their call each and every transmission, this is not necessary and can be annoying. You don't have to give anyone else's call sign at any time, but it is a nice acknowledgment of the other person - like a handshake. In a round table of several folks, it may go on so long that by the time it comes back to you, or an emergency occurs, the ten minutes are up, so best to ID after each of your transmissions when in a round-table.

CTSS FREQUENCIES (Hertz) with Motorola and ICOM Codes (ICOM Codes On Some Older Units)

CTCSS	MOT	ICOM		CTCSS	MOT	ICOM		CTCSS	MOT	ICOM
69.3	WZ	1 = 67.0		110.9	2Z	15		173.8	6A	28
71.9	XA	2		114.8	2A	16		179.9	6B	29
74.4	WA	3		118.8	2B	17		186.2	7Z	30
77.0	XB	4		123.0	3Z	18		192.8	7A	31
79.7	WB	5		127.3	3A	19		203.5	M1	32
82.5	YZ	6		131.8	3B	20		210.7	M2	33
85.4	YA	7		136.5	4Z	21		218.1	M3	34
88.5	YB	8		141.3	4A	22		225.7	M4	35
91.5	ZZ	9		146.2	4B	23		229.1	9Z	
94.8	ZA	10		151.4	5Z	24		233.6	M5	36
97.4	ZB	11		156.7	5A	25		241.8	M6	
100.0	1Z	12		162.2	5B	26		250.3	M7	
103.5	1A	13		167.9	6Z	27		254.1	0Z	
107.2	1B	14								

STANDARD OFFSETS FOR HF/VHF/UHF/SHF

29 MHz	100 kHz	(-)	147 MHz	600 kHz	(+)	1.2 GHz	12 MHz	(-)
50 MHz	500 kHz	(-)	222 MHz	1.6 MHz	(-)	2.4 GHz	20 MHz	(-)
145 MHz	600 kHz	(-)	440 MHz	5.0 MHz	(-) **			
146 MHz	600 kHz	(+ or -) *	900 MHz	25.0 MHz	(-)			

* Lower portion of 146 MHz is often plus offsets, Upper portion is usually minus offsets

** Plus Offset In some areas - notably Northern California

FOLLOWING ARE TYPICAL BAND PLANS – See Area Coordinators http://www.arrl.org/nfcc/coordinators.htm

28 MHz (10 METERS) REPEATER FREQUENCIES
4 Repeater Pairs, 20 kHz Channel Spacing, 100 kHz in/out
29.520 - 29.620   29.540 - 29.640   29.560 - 29.660   29.580 – 29.680

---

50 MHz (6 METERS) REPEATER FREQUENCIES
39 Repeater Pairs, 20 kHz spacing, 1 MHz - in/out
15 Repeater Pairs, 20 kHz spacing, 500 kHz - in/out
For Simplex Frequencies – See Chapter 3

---

144 MHz (2 METERS) REPEATER FREQUENCIES
Repeater Sub-band 144.50 - 145.50 MHz 20 Repeater Pairs, 20 kHz spacing, 600 kHz in/out
Standard Repeater Band 146 - 148 MHz 68 Repeater Pairs, 15 kHz spacing, 600 kHz in/out
For repeater pairs - see Repeater Guides for your area http://ac6v.com/repeaters.htm
For Simplex Frequencies – See Chapter 3

---

220 MHz (1.25 METERS) FREQUENCIES
Repeater Pairs from 222.50 – 224.98 MHz -- 45 Repeater Pairs, 20 kHz channel spacing, 1.6 MHz in/out
For repeater pairs - see Repeater Guides for your area http://ac6v.com/repeaters.htm

For Simplex Frequencies – See Chapter 3

---

440 MHz (70 CENTIMETERS) REPEATER FREQUENCIES
Standard 440 MHz FM Repeaters -- 128 Repeater Pairs, 25 kHz spacing, 5 MHz split, High/input, Low/output
For repeater pairs - see Repeater Guides for your area http://ac6v.com/repeaters.htm
For Simplex Frequencies – See Chapter 3

902-928 MHz (33 CENTIMETERS) FREQUENCIES
20 FM Repeater Pairs
12.5 kHz spacing between channels, 25 MHz Duplex in/out
29 FM Repeater Pairs
100 kHz spacing between channels, 12 MHz Duplex in/out
For Simplex Frequencies – See Chapter 3

1.2 GHz (23 CENTIMETERS) FREQUENCIES
1270.00 - 1276.00 MHz -- FM Repeater inputs, 239 pairs, every 25 kHz
1282.00 - 1288.00 MHz -- FM Repeater outputs paired with 1270-1276 GHz
For Simplex Frequencies – See Chapter 3

NOTES

CHAPTER 5: PROGRAMMING

PROGRAMMING A VHF/UHF RADIO

A common puzzle and frequently heard complaint with the new radio is programming the new goodie. Transceiver programming can be easy and intuitive to ultra complex. The manuals range from very good to awful. This means you are not alone with the problem; engineers, technicians and programmers can have trouble with programming rigs as well. A helpful method is to make a "cheat sheet" of your own, written by you so it can be used by you. A cheat sheet copy in your wallet/purse, at home, and glove compartment will save a lot of aggravation later on and make it a lot easier to program when you travel.

Some radio manufacturers have free on-line manuals with cheat sheets http://ac6v.com/manuals.htm Other sources are the web or the local radio store. These are often referred to by different names, Quick Reference Cards, Radio Setup Guides, Mini-Manuals, etc. N6FN has created a series of high-quality laminated Quick Reference Guides for most of the HTs and mobile radios currently being sold. http://www.niftyaccessories.com/

Maybe best to walk before you run. First tackle the easy one, program a simplex channel. Then try programming for a repeater not requiring PL for access. Next, program for a repeater that does require PL to access (Pled repeater). Then program for a repeater requiring PL and set your receiver for tone squelch.

For a simplex channel, first determine how to access the VFO and put in the desired frequency. Next, find the information for offset. Check that the offset is neither plus or minus, i.e., it is in simplex, then memory write all.

For a repeater requiring a PL tone, invoke the VFO mode and enter the repeater output frequency, set the offset value (e.g., 600 kHz) plus or minus as required. Next enter the tone frequency, tone activation, then memory write all. Many beginners enter the tone frequency, but forget to activate the tone. See Chapter 4 for definitions of VFO, offset, PL, etc. Here is HT programming for a popular model, the ICOM IC-T2H.

1. Push [aV] to get in the VFO mode (as opposed to the memory mode).
2. Press 6 of the digit keys e.g. 1 4 6 7 3 0 to enter the frequency
3. Push [aV] for 1 second - push left arrow one or more times until [oW] appears
4. Push up or down buttons until 0.60 (Mhz) is in the display. Press [aV] twice
5. Push [DUP] key – Press to select either (minus) -DUP or DUP which is a plus offset
6. Push [aV] for 1 second to enter the set mode
7. Push left arrow one or more times until RP appears
8. Push Up or Down key until the desired PL frequency appears e.g., 100 (Hz)
9. Push [aV] twice to set the desired tone and exit the set mode.
10. Push [cT] to activate the subaudible tone. Press sequentially will be:
11. [none], [T], [T SQL (*)], [T SQL] ---- for plain repeater operation, select [T]
12. Push [bM] momentarily (do not hold down), then select a memory channel with the up or down keys.
13. Push [bM] for 1 second to program the information into the selected channel and return to the VFO mode.
14. Push [aV] to get into the memory recall mode. Push up or down arrows for desired channel.
15. Pressing PTT will transmit on the repeater input frequency – releasing PTT – the display should revert back to listen on the output frequency.

Well that is a lot of steps just for basic repeater operation. But if you read the whole manual and write down the steps and make your own cheat sheet, you will be able to program the radio in the future without the manual. Next try your cheat sheet with the manual on the side. Refer to the manual only if needed

If the cheat sheet works without the manual, next start over and program another repeater and channel. Elaborate on the cheat sheet as necessary. Repeat programming channels until you are absolutely certain the cheat sheet will work for you without flipping pages in the manual. Only when you can go thru the entire process without the manual and only using the cheat sheet – then by George, I think you got it.

Why is all this necessary? Well in my HT manual, VFO activation is on page 11. Offset is on page 16. PL tone is on page 17. Page 15 to activate the tone function. And finally page 19 to get it all in a memory channel !!! There is a lot more to programming, other functions such as the call channel, lock, setting tuning steps, auto-repeater function, scan mode, tone squelch, and lots more. Best to read the entire manual and make a cheat sheet for all the functions you intend to use.

If you use only one repeater most of the time and are prone to inadvertently hitting keys or mic buttons – put the unit in lock to prevent going off channel or changing other functions. The auto repeater function can be activated or disabled. When activated, the standard repeater information is in effect for the various band portions, i.e., duplex direction plus or minus, or off, tone encoder on or off.

Scan modes are convenient for scanning all frequencies or just the memory channels, the first when traveling, the latter for roaming around a given area. Some of the new multi band units allow programming of AM and FM radio frequencies, WWV, Shortwave frequencies, as well as NOAA Weather announcements. Some units allow groupings, to separate AM/FM/SWL from the Ham bands. In effect it can function as a scanner. Note that programming allows for scanning ranges, such as band edge to band edge or a portion thereof.

Since most modern rigs have 100+ memories, some operators program in all the USA VHF/UHF repeater pairs and simplex channels. For a list of these see URL: http://www.bloomington.in.us/~wh2t/two.html

One can do pre-programming prior to a trip – USA or overseas. I successfully operated in Europe under the CEPT agreement URL: http://www.arrl.org/FandES/field/regulations/io/ In some countries you must have CEPT papers or a temporary license else your equipment may be confiscated. See operating abroad http://ac6v.com/dxpeditions.htm#LIC Before using the CEPT agreement, be sure to read the proper way to ID and the requirements for paper work you must carry. A rudimentary knowledge of the language will help, as VHFers may be less conversant in English than their HF counterparts. Some countries use tone burst rather than PL and these can be programmed into many HTs. Also the offsets in some countries are different than the USA. Most modern rigs will accommodate the various offsets and tones. See World Repeaters – URL: ac6v.com/repeaters.htm

Last but not least if the programming boggles the mind (it does mine at times), consider buying the programming cable and software. Some manufacturers have the software free on their web site. Using a computer, programming cable, and software makes programming a breeze. Groups can easily be down loaded into the HT or mobile radio from the computer. Some units allow names rather than frequencies to be displayed. An example is PARC (Palomar Amateur Radio Club) can be displayed instead of 146.730. This is a big help if you have programmed in a lot of frequencies.

CHAPTER 6
ANTENNAS, POWER SOURCES, VSWR, & DECIBELS

BETTER ANTENNAS

Raising power will get you a better signal into the repeater, but if you are operating an HT with battery power and using full transmit power, the battery drain is pretty severe, to say nothing of heating up your hand. Better to look at an improved antenna. Today's HT's are indeed a wonder, but a rubber duck is a pretty lousy antenna -- designed for convenience, not performance.

Tests show the rubber duck has about minus 5db gain (see decibels below) compared to a quarter wave antenna held at shoulder height. In terms of effective radiated power (ERP), a 5-Watt HT with a rubber duck antenna, held at shoulder height radiates an effective power of about 1.5 watts. Placing the HT on your belt could attenuate the signal another 20db, reducing ERP to only 15 milliwatts! UHF results were even worse.

A vast improvement can be made by using a 1/4 wave 1/2 wave, or 5/8 wave after market antenna for your HT. These are made by MFJ, Smiley, Diamond, Comet and others. Ask at the local Radio Store. But with "micro HT's" a big 5/8 wave antenna (~ 47.5" fully extended), may be a case of the tail wagging the dog - putting undue stress on the little HT connector -- maybe try before you buy.

Or you can add a Tiger Tail which for 2 meters is simply a 19 1/4" piece of wire connected to the outer barrel ring of the rubber duck antenna. MFJ makes a twin lead J-Pole (model 1730) that works well in lieu of a rubber duck.

For mobiles, a mag mount or permanent trunk-lid or roof mount will greatly improve your HT operating range but may result in intermod. Glass mount antennas work well - but some autos have a tinted passivated (metal particles) wind shields -- ask your auto dealer. Mobile antennas can be used for the base station as well, usually requiring a homebrew ground plane of some kind -- ask your elmer. I use a baking sheet for a mag mount antenna used indoors. Also homebrew antennas for your base station can be built -- see Antenna Projects. http://ac6v.com/antprojects.htm And of course commercial base and mobile antennas can be used -- see http://ac6v.com/antdealer.htm Adjusting antenna length and VSWR are discussed below.

Vertical antennas are commonly used for repeater work as repeater antennas are almost always vertically polarized. Example, FM commercial broadcast antennas are vertical and so is your car antenna. You can use horizontal yagis or halos but severe cross polarization attenuation (up to 20 or 30db) can result when the other station or repeater has a vertically polarized antenna. However one can configure a beam antenna for vertical polarization, just mount the beam vertically or feed a quad at the sides. Most use a non-conducting mast when mounting a yagi vertically.

Horizontal antennas on VHF/UHF are more commonly used when operating SSB or CW. For repeater work verticals are desirable because of polarization and they are omni-directional. You may want to work repeaters in several directions using a directional antenna. Beams of course require a rotor.

Popular vertical antennas are 1/4 wave, 3/8 wave, 1/2 wave, 5/8 wave and J-Poles. The 1/4 wave, 1/2 wave, 3/8 wave, 5/8 wave, and J-Poles give gain over a reference antenna called an isotropic radiator (dBi). Sometimes manufacturers give gain referenced to a dipole dBd. There is about 2.1 dB difference between the isotropic and the dipole. An isotropic is a theoretical antenna that radiates in all planes of a sphere. Gain, of course is squishing radiated energy into a narrower pattern, and it increases your effective radiation power (ERP). You can't get out more than you put in, but a directional antenna will concentrate the power into a narrower pattern, hence "gain" compared to a reference antenna. For verticals the pattern is squished into a lower radiation angle but is still omni directional. The receive pattern of an antenna is the same pattern as when the antenna is transmitting, that is reciprocal.

Gain can also be achieved with co-linear elements. That is, one vertical element stacked above the other, inline. **More important is height**, which can make a substantial difference in perceived "gain." VHF/UHF is principally limited to line-of-sight transmissions - hence the advice for height. The higher you are, the further you can see.

Below figure 1A shows a vertical antenna radiating at a moderately high angle. Figure 1B and 1C show "gain" vertical antennas radiating at lower angles. Keep in mind that the antennas are radiating 360 degrees (omni directional) se e Figure 2. With the gain antennas, the angle of radiation has been changed and the lobe has more power in the desired direction. However antenna 1C has developed another lobe at a fairly high angle, this can occur if the antenna is greater than 9/16 wavelength

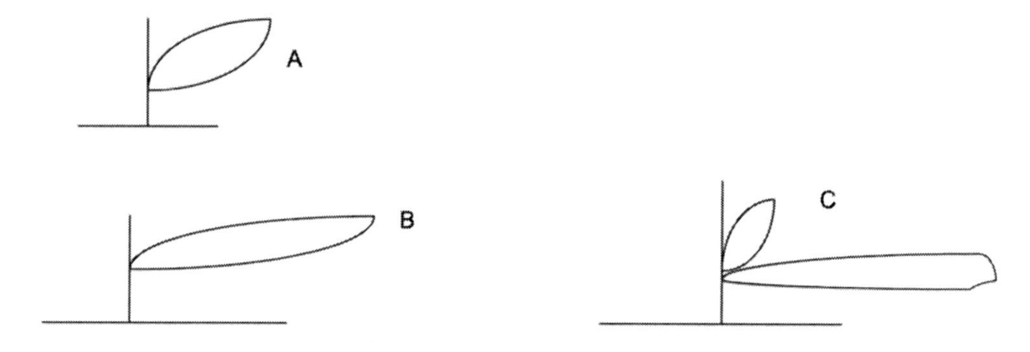

Figure 1. Vertical Antennas patterns

The antenna with the higher radiation angle might work well when down in a valley when trying to get into a repeater high up on a mountaintop. The low angle radiators may work well for long-range communications, but might also have sufficient signal for local repeaters.

A quarter wave antenna is easy to build or buy and is easy to match to the feedline. The radiation angle is higher than the others mentioned, but this may be fine where you are in a valley situation and/or the mountain top repeaters are nearby.

The 1/2 wave, 3/8 wave, and 5/8 wave antennas require impedance matching networks but will have progressively lower radiation angles which works well for distance communication. The 5/8 wave is somewhat the optimum for low radiation angles, the real optimum is 9/16 wave – above that and more the patterns start to have split lobes (Figure 1C) and may be counter-productive. Although the 5/8 wave will have some small lobe splitting, it is frequently used because of its ease of construction and matching. I have both 1/4 wave and 5/8 wave whips for my car using a roof mount.

Around town I use the 1/4 wave and I can get in and out of the garage without scraping up stuff. The 5/8 wave is great when traveling, enabling me to get into repeaters well down the road. A tilt mount can be installed for tall car antennas.

J-Poles are used for base stations and are something of a plumbing project usually using copper tubing commonly available at hardware stores. For all practical purposes, it has the same gain as a ½-wave dipole and does not require a ground plane. See Antenna Projects URL: http://ac6v.com/antprojects.htm

For base stations, yagis, quads and other directional antennas will give significant gain in the forward direction. They can be configured for horizontal or vertical polarization. Directional antennas are the choice of Dxers, contesters, CW and SSB folks, or FM users at extreme distances from repeaters.

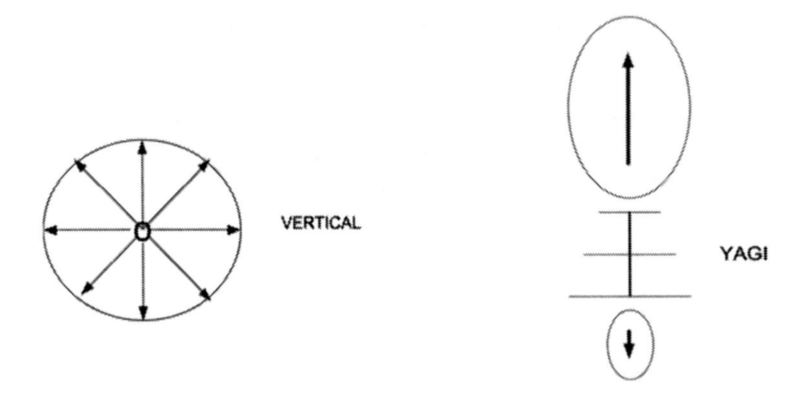

Figure 2. Vertical Radiation Patterns vs. a Directional Antenna

Looking at antennas from above, the vertical antenna on the left radiates equally in all directions (omni-directional) however with the lobes as shown in Figure 1. The beam antenna at the right concentrates energy in the forward direction, perhaps 6 dB gain. 6 dB gain means quadrupling the effective radiating power, e.g., 20 Watts to 80 Watts is 6 dB. Note that it has a back lobe as well that is considerably down in energy perhaps 15 to 20 dB. The lobes shown are at the half power points (3 dB) so there will be lots of energy outside the lobe itself. The front to back ratio can be very effective in receiving signals from the desired direction and minimizing others. With yagis, a big null occurs at 90 degrees away from the main lobe.

In addition to the antennas already discussed, others are: dual band, tri-band and quad-band antennas to cover several bands. Log periodic directional beams are very wide band, e.g., frequency range 26 to 1300 MHz continuously, boom length 12 feet, 24 elements, weight 12 pounds, average gain 6 dBi. A discone antenna also has a wide continuous frequency range e.g., 60-1000MHz and is mostly omni directional, however gain is low. Halo antennas are horizontally polarized and have gain of about 6 dB over a quarter wave. Delta loops, quads or quagis can be built; a quagi is part quad and part yagi and work well at the VHF and UHF frequencies. Collinears are stacked antennas, which can achieve up to 9dB gain.

Another way to look at antennas is with a light bulb analogy. Consider a light bulb that shines equally in all directions, this is analogous to an isotropic radiator. Putting a reflector behind the bulb (like a flashlight) redirects the back energy – forward to reinforce the forward light and we get a beam.

Obviously the beam has increased in brightness compared to the plain bulb. We can also direct the flashlight energy in different directions as we can with a beam antenna.

Another analogy is a toy balloon. When perfectly round – an isotropic. When squished, we can form a big "beam" in a desired direction. If squished just right a little beam will squirt out the back end.

When building and installing antennas, an Antenna Analyzer such as those made by MFJ are invaluable. Or at least get hold of an SWR/power meter to check for the resonance point of the antenna and assess proper impedance matching. Solid state rigs abhor swrs above 2.5:1 so best beg, borrow, or buy one for VHF/UHF antenna installation to ensure proper operation. Be aware that some VSWR meters are for HF use only, check the specs for frequency ranges.

Base antennas should be mounted up as high as possible clear of obstructions. However base antennas can be installed indoors if that is all you can do. Contrary to popular belief, RF will go through walls and stucco chicken wire etc., and magically get to a repeater or a simplex station. Perhaps not as well, but a lot depends on distance and repeater location altitude. I have indoor VHF/UHF antennas and get into local repeaters quite well. One repeater I access is 90 miles away but the path to it is over the ocean.

With today's antenna restrictions and homeowners associations, stealth antennas are in order, see URL http://ac6v.com/antprojects.htm#STANT

An excellent way for HT owners to upgrade their portable emergency equipment is to obtain a power amplifier (jargon - brick). An ideal amp should weigh less than 1 pound, be capable of 10 to 15 Watts output when driven by an HT at 1 Watt, or 20 to 40 Watts output when driven by 2 to 3 Watts. Current draw should be no more than 8 amperes at its maximum rated output, enabling it to operate safely from the .093-diameter pin Molex Series 1545 connector or possibly a fused cigarette lighter plug.

VSWR (Voltage Standing Wave Ratio)

Modern solid state radios hate high VSWRs and you must check the VSWR when installing antennas. VSWR is an indication of forward and reflected power. Reflected power is undesirable and results from a mismatch in impedance between the transmitter and the antenna. If the antenna is too long, it will resonate at a lower frequency than desired, conversely if the antenna is too short, it will resonate at a higher frequency than desired, analogy - like short and long guitar strings. In either case, the VSWR may be too high to operate on the desired band, so some length adjustment may be required. Some whip antennas are supplied quite long and you will need to cut the antennas until a low VSWR is achieved over the range you wish to operate. Antenna manufacturers supply a sheet showing lengths versus frequency, but you should still use a meter or analyzer to be sure the system is matched properly. Easy to shorten, but not lengthen, so chop off a little at a time. Chop a little, measure VSWR over the range and repeat.

A VSWR of 1:1 is the theoretical ideal but unfortunately cannot be achieved. This value would indicate 100% efficiency in a transmitter, transmission line and antenna system, with no reflected power. But with proper matching, VSWRs can be adjusted to approach the ideal. Most modern transceivers detect a poor VSWR and reduce the output power in order to keep the reverse power down to a safe level. VSWRs of 1:1.5 to 1:2.5 are generally considered acceptable values. If this is too scary (and it can be), ask an elmer who has the equipment and know how to help you set it up. To check VSWR, use an SWR meter or antenna analyzer, the manuals will give instructions as how to set up the test.

Theory of VSWR gets into antenna design and you may want to read all about VSWR at URL: http://www.antennex.com/preview/vswr.htm

Following is a list of great web pages for antenna theory and dimension calculators. See URL: http://ac6v.com/antprojects.htm#theory Also hundreds of homebrew antenna projects can be found there.

ANTENNA THEORY AND BASICS

Antenna Basics and Theory
Antenna Dimensions Calculators
Antenna Gain
Antenna Tuner Theory
Baluns
Baluns and Choke Baluns
Choosing Wire For An Antenna
Dipole End Insulator Installation
Height Of Dipoles
Tower Specs, Advice and Tips
Transmission Lines – Open and Coax
Vertical Antennas
VSWR - Decibels Too

DECIBELS

The decibel, (1/10 Bel) or dB, is a means of expressing the gain of an active device (such as an amplifier) or the loss in a passive device (such as an attenuator or length of cable). It is simply the ratio of output to input expressed in logarithmic form. It is also used to express the gain of an antenna relative to a reference antenna. Without getting too far into the math, here are some handy easy to remember rules of thumb.

A power ratio of two is equal to 3 dB. Conversely, a power ratio of 1/2 is equal to -3 dB. (NOTE: The -3 dB point is often referred to as a "half-power point," as when describing a frequency response curve or an antenna lobe pattern. Following is the math for doubling power from 10 Watts to 20 Watts:

dB = 10 times $\log_{10}$ P1/P2; thus dB = 10 times $\log_{10}$ 20 Watts/10 Watts; so dB = 10 times $\log_{10}$ of 2 The logarithm of 2 is 0.301 and 10 times this is 3.01 dB, for all practical purposes 3 dB.

A power ratio of 10 is equal to 10 dB. Conversely, a power ratio of 1/10 is equal to -10 dB.

Example of use: Let's raise the power of a transmitter from 10 Watts to 160 Watts. 10 watts to 20 watts is 3 db. 20 watts to 40 watts is 3 dB. 40 to 80 watts is 3 dB. 80 watts to 160 Watts is 3 dB. Since dB's can be added, an increase of 10 watts to 160 watts is 12 dB (3 dB +3 dB +3 dB +3 dB). One receiver S-unit is 6dB, thus raising power from 10 Watts to 160 Watts at the transmitter will only result in two S-Units increase at a receiver. The 6 dB per S-unit is not an industry standard but is commonly quoted. Brands vary.

BATTERIES

It is almost a necessity to have two battery packs. One for immediate use and the other in the charger. There is lots of misinformation, myths, and other bad advice regarding the care and feeding of HT batteries. For the straight scoop, read all about batteries at URL http://ac6v.com/techref.htm#BATT

Unfortunately, battery packs usually have the cells in series. So if a one cell is weak it may reverse and go goofy when the battery gets down toward fully discharged. Conversely, charging problems may occur if a cell is weak. Some manufacturers do a good job of matching the cells, whether all manufacturers do is questionable.

When the cells are matched they all charge and discharge in unison and long battery life can be expected. This may explain why some battery packs last a long time while others die pretty quickly. But those that die may have been mistreated along the way.

Sealed Lead Acid (SLA). These batteries use the same technology as your car battery. They are relatively inexpensive, and provide 200-500 recharge cycles depending on usage. SLA battery packs give high capacity but are fairly heavy, and are more sensitive to temperature and deep discharge/overcharge than other batteries. SLA batteries should be stored in their charged state as leaving them uncharged for long periods can cause permanent damage. Some very high capacity battery packs are available and can be easily adapted to Amateur transceivers - such as a 6 amp-hour battery used for car models.

Alkaline Batteries. These are great backups if your HT will accommodate them. Some HTs can be purchased with a shell that allows standard AA batteries to be used.

When hiking or traveling, and away from being able to recharge the battery pack, alkalines are a great back up to allow continued operation. However, they do have a sloping discharge characteristic, unlike some other battery types. Be sure to use the same brand and match the set.

They have extremely long shelf life which can be up to 2.5 years and typically have a storage discharge rate of 0.01% per day. Rechargeable batteries vary depending on type but a shelf life of 60 days at a discharge rate of 0.5% per day is typical for rechargeables.

AA batteries have a capacity of about 2,000 mAmp/hours but the battery capacity will be better with lower drain currents. A myth is that rechargeable batteries have a lower capacity than disposable alkaline batteries. This really depends on the battery application. For most high drain electronic devices, like digital cameras for example, rechargeable batteries will continue to work much longer than alkaline batteries. In fact in devices like digital cameras, NiMH batteries will run on a single charge for 3-4 times as long as they would on an alkaline battery.

Nickel Cadmium (NiCad). Given equal power output and run time, NiCads are about 25% lighter than lead-acids. You can expect around 700 or more recharge cycles depending on usage.
They don't mind deep discharge, and perform better than SLAs at low temperature. Contrary to urban legend, today's industrial grade NiCad batteries do not suffer appreciably from memory recharge problems. Because of a weak cell possibility, NiCads should not be discharged to the dead radio point, more specifically manufacturers recommend -- don't discharge below 1.0V per cell.

While NiCads have many advantages, be aware that towards the end of their power supply, NiCads go from good to near zero rather abruptly (a rude surprise if you are in a sticky situation).
A safe way to charge NiCads is at 1/10 of their capacity overnight. For slow charging, a 1000 ma/hr battery would be charged at 100 ma. A constant current source is a good way to charge NiCads. Fast chargers can be used if properly designed and have temperature monitoring. Some use an impulse current method. NiCads are considered completely charged at 1.43V per cell.

Nickel-Metal Hydride (NiMH). Their performance is similar to NiCads, but they provide a power to weight savings of 25 to 40% versus a comparable NiCad battery. They feature high energy density, flat discharge, very high rate performance, and fast charge capability. Moreover, they are much less toxic during the recycling process. Unfortunately, NiMH batteries are fairly expensive. For details of Charging NiMH batteries – see URL http://www.powerpacks-uk.com/Charging%20NiMh%20Batteries.htm

NiMH and NiCd batteries self discharge at a MUCH faster rate than alkaline batteries. In fact, at "room temperature" (about 70 degrees F) NiMH and NiCD batteries will self discharge a few percent per day. Storing them at lower temperatures will slow their self-discharge rate dramatically. NiMH batteries stored at freezing will retain over 90% of their charge for a full month. So it might make sense to store them in a freezer. If you do, it's best to bring them back to room temperature before using them. Even if you don't freeze your NiMH batteries after charging them, you should store them in a cool place to minimize their self-discharge.

Lithium-ion Batteries. Lithium-ion batteries have several advantages: They have a higher energy density than most other types of rechargeables.

This means that for their size or weight they can store more energy than other rechargeable batteries. They feature excellent rate capability and low temperature performance; fair shelf life; and a relatively flat discharge characteristic.

They also operate at higher voltages than other rechargeables, typically about 3.7 volts for lithium-ion vs. 1.2 volts for NiMH or NiCd. Lithium-ion batteries will retain most of their charge even after months of storage. See URL: http://www.powerstream.com/li.htm for charging characteristics.

Auto Power Accessories. Some HT accessories may allow the HT to operate from the cigarette lighter plug and also accessories may be available to charge the HT from the same plug.

About Capacity. Assigning capacities to batteries can be very tricky, that's probably why you don't see capacity ratings marked on most alkaline batteries. When powering high drain electronic devices like transceivers, digital cameras, etc., an alkaline battery will only deliver a small fraction of its rated capacity.

A NiMH or NiCd battery is likely to deliver much closer to its rated capacity when it is powering high drain devices. This means that a NiMH battery with a rated capacity of 1800 mAh can take many more photos than an alkaline battery with a rated capacity of 2,800 mAh!!

For gel cells, a 20 amp hour battery will give 1 ampere for 20 hours (referred to as the 20C rate), but will only give 20 amperes for about 60% of an hour or 36 minutes at which point the battery will appear to be discharged. Over time, a useful charge may return.

The reason for this battery recovery is that some unused electrons have had time to migrate from the center of the plates to the edge of the plates at the lead/acid interface where the reaction occurs, so the battery has come "back to life". URL: http://www.hamcontact.com/products.html

DC POWER SUPPLIES

A mobile or an HT can be operated in a base situation by using a DC power supply to convert 117VAC to the level required by the radio. Many are available with a 12VDC output to directly power a mobile radio. Powering an HT from a 12-Volt supply will require building a regulator circuit to achieve the proper voltage. Be sure and check your manual for the maximum voltage input and current requirements.

Wall modules may work but be sure to check that they have the necessary current rating and provide adequate filtering. Note that the wall module supplied with the HT is for charging only (unless the manufacturer specifies otherwise) – they will not support transmitting– not enough current. If you try to transmit, they may work but with excessive AC ripple (hum on your signal). And it may overheat the wall module and frazzle it. Adequate filtering means the power supply is giving near perfect direct current without appreciable AC ripple. Filters can be built with inductors and capacitors.

A neat emergency supply is a high capacity gel-cell battery with a handy carrying case. Some of these have meters and selectable voltage outputs. An example is the Power Station rated at 7 Ampere-Hours. It features a 12V cigarette lighter output. Also a 1/8" jack on the side provides 3, 6, or 9 VDC. Chargers are provided for charging the supply from the house or a car (both with auto shutoff). See URL: http://www.hamcontact.com/products.html Smaller ones are available with a shoulder carrying strap or fanny pack.

Another oft-used idea is a regular car battery for use with a mobile radio. A better choice is the RV style deep-cycle batteries. A standard battery charger can be used to charge the battery after use and put it on trickle charge when not used. This will last for several days under normal or emergency use. And of course in a dire emergency, you can use your car battery and recharge on engine power if need be. And solar power can be used to charge batteries. A good back up for HTs is several alkaline packs. Use appropriate connectors to make the interface and observe both polarity and voltage. You are out of AAs – no you aren't – go raid the remote controls, clocks, toys, etc.

AC GENERATORS

To really get into portable power supplies, gasoline generators that supply 120 Volts AC and 12 Volts DC are available at various wattage ratings. These can be used to power home devices as well as Ham gear. Hams use these during Emergencies, Field Days, Special Event stations, and Dxpeditions.

Typical generator specifications are: 3.5 HP; AC Rated Output - 120V, 1600W, 13.3A; DC Output - 12V, 96W, 8A; Fuel Capacity 1.1 Gallons; Run Time 4 hours rated load, 15 Hours a one fourth load; Dry Weight 46 pounds; Carry Around Handle. And lots of gasoline in your auto should you need fuel. Have a siphon?

CHAPTER 7: USING REPEATERS

FINDING REPEATERS

One can set their receiver to scan the band and listen for repeaters, but a repeater may not be transmitting or seldom used, so it is catch as catch can. To find repeaters in your area, ARRL repeater guides are published and guides are also available free on the Internet -- see Web Repeater Guides. http://ac6v.com/repeaters.htm

Many are open repeaters -- open to all licensed hams. Others are closed for members only and require membership. Some are specialty repeaters such as Emergency or DX repeaters and not amenable to other uses. DX repeaters are used to disseminate HF DX information. The repeater guides usually specify this. Not specified is that some repeaters are primarily "ragchew", others are "calling" repeaters, and some are general purpose. Listening for a while will usually reveal the repeater's primary use or you can always ask on the repeater for repeater usage guidelines. In large metropolitan areas, coverage of repeaters sometimes overlap and it is hard to avoid hearing two repeaters in certain locations. Tone Squelch may help in some cases.

Most radios offer a Scan function that allows repeater bands to be scanned. Either VFO (step-continuous) or memory scan can be used. When a signal is encountered, the radio either pauses for a bit or stops on the encountered signal (see manual for options). Using PL seek (if available) will allow you to lock on the PL frequency used by the repeater. This is great for traveling when you don't have a repeater guide or are too busy driving to fool with radio settings.

Since modern radios have in excess of 100 channel memories, some users will program their radios for all standard repeater pairs. For typical HF/VHF/SHF Standard pairs, see Chapter 4 and URL: http://ac6v.com/repeaters.htm

For the traveling ham, USA Repeaters and indeed World Wide Repeaters, from Austria to Turkey, from Alaska to Wyoming are listed at http://ac6v.com/repeaters.htm#WWW Also the ARRL has a CD ROM -- TravelPlus for Repeaters™ CD-ROM . Includes User Definable Maps. Repeater List Advanced Search. GPS Tracking Automatic List Refresh – Automatic list display of all repeaters within defined range during GPS Tracking. http://www.arrl.org/catalog/travelplus/

HOW DO WE USE REPEATERS?

Perhaps after reading the following guidelines to protocol, it appears too formal or restrictive. This is not the intent, the whole purpose of this book is to acquaint the reader with what goes on - on a repeater - terms, jargon, and how a repeater works from a user stand point. There are literally thousands of repeaters across the US (and the world). Each one can have its own peculiarities and unique operating procedures, but there are some basics that apply to many of them. Be aware that some repeaters are very informal and just about anything goes within the FCC rules. Others are very formal – best listen in for a while.

Still others have several hundred users as well as visitors and to create order out of chaos, these repeaters are more organized as to protocol and usage. Act accordingly. Use Plain Language. Talk Like You Talk On The Telephone. You can pick up the jargon as you go along.

If you are new to the repeater scene, maybe LISTEN for a while to catch on to the swing of things. Repeaters are party lines. Lots of people use them on and off throughout the day, and the one you've selected may be busy with another conversation. So listen for a minute or two. If you are in a hurry and not sure if the repeater is in use, just give your call sign and ask - "AC6V - Is the repeater in use?" This will avoid interrupting a conversation (QSO) or an emergency. Throughout this chapter, I'll use my call sign in the examples, simply substitute your call for mine in the examples.

If the repeater is quiet, key your transmitter and announce "AC6V listening", this is short hand for "AC6V listening for any call". Or "AC6V listening - anyone on frequency?" Some just announce simply "AC6V". Others give an idea of their situation - "AC6V Mobile", "AC6V Maritime Mobile". Although not as common today, "AC6V monitoring" was at one time used by control operators and in some areas may still be the case, but this seems to have given way to where monitoring and listening are now synonymous.

Don't call CQ on repeaters (HF Band Stuff) -- many VHFers don't savvy it and just isn't used on FM repeaters (on VHF SSB OR CW - Yes). If you insist on calling CQ on a repeater, it is not against the rules – but 16 old time VHF hams will come back to you with the lecture "We don't use CQ on repeaters".

If somebody answers, you have a contact! Give your callsign phonetically (Alpha Charlie Six Victor) when making new contacts so they can copy clearly and get your callsign correctly. A conversation follows. If you have a contact, when you hear the beep and no else comes in -- it is your turn to talk.

If no one answers, perhaps no one is on channel, but more likely they are busy, not inclined to talk, building something, soldering, etc., not ignoring you or being rude, but not responding for their own reasons. Local repeaters across country may not care to "entertain" travelers but will usually answer questions about the area. Repeating your "AC6V listening" a few times is OK, but repeating it several times is probably a waste of air space. There is no need to acknowledge a negative contact, but many do as a matter of letting others know they are clear of the repeater (AC6V clear). When calling another station e.g., "WN6K - AC6V Calling" and there is no answer, then perhaps finish with "Negative contact AC6V clear (or listening)".

If you want a quick "Radio Check" to see if you are accessing the repeater adequately -- try "AC6V radio check please" or "AC6V signal report please" will usually solicit a come back but not necessarily an extended conversation. Another way to solicit a conversation is to ask a pointed question, "AC6V - anyone on frequency tell me the PL for repeater X?" "AC6V – Can some one tell me how the traffic is on I-15?" But don't key up (kerchunk) with out Identifying (IDing) – it's illegal and annoying. See Kerchunking below.

Since repeaters use Frequency Modulation (noise free) and are channelized, it is conducive to using plain language, as you would do on a telephone. Calling CQ, using Q-signals, and HF jargon are not necessary although there is much jargon that is commonly used as well as a few Q-Signals. These will be discussed later so you will recognize them. Keep in mind that some VHF/UHF users have never been on HF and are not familiar with the terms.

You can talk about almost anything you want -- there are few rules about the content of Amateur conversations. You can't use Ham Radio to do business, but you can talk about where you work and what you do. It seems so easy to autopatch into work and pick up a message, tell the boss you're going to be late, etc, but these work type messages are a no no. Remember, the FCC prohibits all business communications on Amateur Radio (except in emergencies).

Unlicensed persons may participate in amateur radio communications, provided a control operator identifies at the beginning and end of transmission, and is present and continuously monitors and supervises the communication to ensure compliance with the rules.

Prime time TV language has been peppered with some incredible obscenities, and so has language on some repeaters. Most repeaters discourage or prohibit this as sensitive folks and children may be listening. Indeed obscenities are against the FCC rules. You're not having a private conversation -- you have lots of listeners, some of them women and children. Non-Hams may be listening on scanners – let's keep an image to the public of sounding professional. Keep that in mind as you choose language and subject matter. Also until you get to know folks, controversial subjects such as religion and politics can get some nasty replies, depending on how open or closed-minded the other person is.

Be aware that some repeaters have a recorder and the audio track is recorded at all times or others will record all autopatch conversations. The voice logs are reviewed daily. Inappropriate language use or meanings will invoke a cautionary comment from the repeater trustee. Continued abuse may draw legal action. The log (files) will be used as legal evidence against the offender if required.

As to what is talked about -- it covers just about anything folks talk about. Technical subjects such as SWR, Antennas, Latest radios, Theory. Operating such as Emergency Procedures and Drills. DX. Contests. History. Dirigibles. Old Cars. Pick Up The Dry Cleaning. Bring Home Some Lettuce. Old Timers reminisces. Old Radios. Well you get the idea and some interesting conversations do take place. But starting a conversation about politics or abortion with a stranger can get dicey .

How long do you talk in view of the party-line concept? Usually there are no hard and fast rules. It depends on the time of day (rush hours are prime time for mobiles, evening is also a busy time, while 2 a.m. is pretty quiet). If you've been interrupted several times by others needing the repeater to call someone, maybe you've been on a bit too long.

Also if you are going to give a long dissertation, let the repeater drop occasionally so as to not time out the repeater and let others standing by to make a quick call. Announce this by "Reset", "Let me Reset" "Let me get another nickel" "Let me burp the repeater" and other cuties. Unless the repeater is very quiet, long dissertations probably should be taken to simplex or a quiet repeater.

Speaking of cuties, CB is great Americana and has some colorful expressions, but these are not often heard in Ham Radio. Best leave these on the 11 Meter band as some old time Hams get all upset with CB expressions -- their problem, not yours --- but just so you know! 10-4 Good Buddy, the first personal is, hitting me with 10 pounds, passing you the good numbers, on the side, come on back etc., best be left on the CB Band. Note that some repeaters have hundreds of users and may have some pretty tight guidelines or even rules regarding usage. For example, a repeater in Los Angeles discourages base station usage during commute hours. Be sure and get the repeater guidelines or lacking same, ask on the repeater if it is open to any users, members only, commute hour guidelines etc.

Generally, repeater use priorities are in order; emergency traffic, nets, autopatches, mobiles and portables, and finally base stations. A repeater in farmland USA may discuss corn and hog prices et. al., and not be amenable to "entertaining travelers". Some use DCS and won't hear you at all. When you finish your transmission and unkey, the beep will alert the other person that it is their turn, so no need to say over, back to you, go, your callsign - their callsign, come on back, etc.

When you un-key your transmitter, most repeaters will stay on the air for a few seconds (hang time), and many will send some kind of "beep or courtesy tone." Then after the hang time, the repeater transmitter drops off the air. **The beep (if so equipped) is there to remind everyone to leave a pause between transmissions in case someone wants to break in.** Even if there's no beep, leave a pause. Somebody may have just come across a traffic accident and needs the repeater to report it. In addition the beep resets a timer. See Timers Below. Some systems use a two tone, bubble up tone, or be-bop tone as a courtesy tone, which may also signal various conditions of the repeater or an inter-tie system. See your repeater guidelines and Chapter 4.

Following the beep, a brief delay is heard before the repeater drops out. Below is a TYPICAL repeater-timing diagram, however PLEASE NOTE that various repeaters have variations on this and the times.

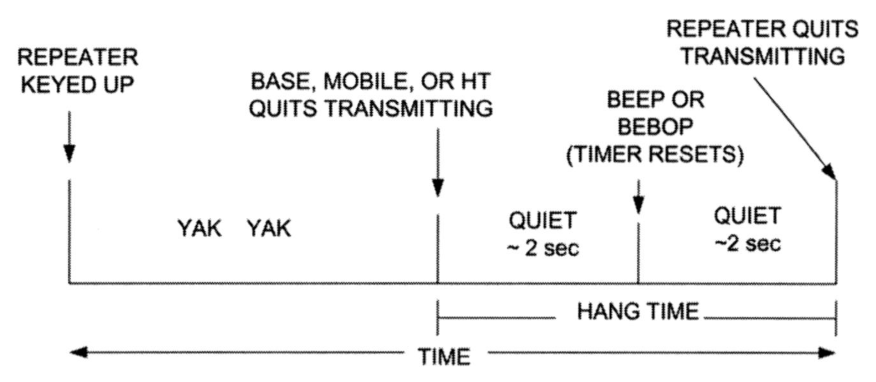

How long do we pause after the beep? You can start talking after the courtesy tone since the timer is reset, but many leave two or three seconds after the repeater drops out, long enough for timid or uninformed folks to come in. 10 to 30 seconds delay leaves everyone wondering why you disappeared or someone

just tuning in listens for 10 seconds or so, hears nothing and comes in - makes a call -- the other party answers and the whole thing gets screwed up.

If you talk too long, a timer times out and the repeater drops off the air. Timers (sometimes called a QSO timer) are typically 1 to 3 minutes depending on the machine. QSO is a Q-Signal (Chapter 8) - meaning conversation. To signal a timer dropout, some repeaters use a voice announcement, others use a series of short beeps, such as a triple beep. Others may use a short delay, a short beep and then dropout.

After timeout, typically there may be a several second wait before the repeater can be brought back up on the air. If the offending talker or a carrier persists, the repeater will remain mute until the input signal disappears, so timeouts can be quite long. Typically, the repeater will ID when it comes back on the air. Note that transmission times are accumulative if folks don't wait for the beep.

Example: A machine has a 60 second timer and station X talks for 50 seconds, if station Y comes in before the beep, 10 seconds later the timer will shut down the repeater.

REPEATER PRIORITIES

Most repeaters consider the repeater priority use as: 1 Emergency and priority traffic. 2 System tests and maintenance . 3 Public service. 4 Nets. 5 Mobile and portable use. 6 Fixed (base) station use.

WE PAUSE FOR STATION IDENTIFICATION

The RULES say YOU must ID once every 10 minutes and when you sign off. Also best to give your callsign when you first come on so folks know who you are. No need to overdo it however, some folks give their call each and every transmission, this is not necessary and can be annoying. It is not necessary to give your callsign after each transmission -- only necessary each ten minutes to satisfy FCC rules. You don't have to give anyone else's call sign at any time, but it is a nice acknowledgment of the other person - like a handshake. In a round table of several folks, it may go on so long that by the time it comes back to you, or an emergency occurs, the ten minutes are up, so best to ID after each of your transmissions.

KERCHUNKING – Unfortunately many kerchunks can be heard on repeaters. The repeater comes up, no one says anything and it drops. Someone is testing to see if they are accessing the repeater. This is illegal for one thing, annoying for another. Better just say "AC6V Testing, No Response Necessary". Perhaps the tester is making an SWR measurement, checking a new installation, or deep down in the innards of the radio tweaking an adjustment, so they don't want to rag chew at the moment, so probably would appreciate no answer, otherwise they would ask for a radio check.

Some repeaters have an anti-kerchunk feature; this requires a carrier with audio before the machine will come up. Some anti-kerchunk filters have an undesirable side effect of losing the first quarter-second or so of audio when the transmitter first keys up. So you need to hold your PTT button down for about 1/2 second before you start talking to make sure the first syllable or so of what you say isn't lost in the kerchunk filter.

GREETING OUT OF AREA VISITORS AND HELPING HAMS WITH DIRECTIONS

Many visiting and vacationing Hams will bring their mobile or HT radios and use the repeaters to find their way around and find attractions. We really need to greet these folks in the spirit of Ham radio – after all we will expect the same when we travel.

Visitors often use their portable designator e.g., AC6V portable 2 when I am in New York or New Jersey. Often heard on repeaters is folks asking for driving directions. Base stations often have access to the Internet and can bring up map quest and do a good job. First determine the location of the mobile before giving them your navigation expertise.

Unfortunately, one person may give adequate directions, and then someone else JUST has to break in to give his/her favorite back alley secret shortcut. Or some other clown gives a two-minute dissertation of street names and landmarks, ad-nauseum.

The lost ham thanks everyone, pulls the plug, and finds a service station to de-confuse themselves. If someone has given adequate directions, let it alone. Make corrections only if the directions are dead wrong. Keep directions short and simple. If one asks for your QTH, they are asking for your location.

PLANES, TRAINS AND AUTOMOBILES – SHIPS TOO.

For those that want to operate on any transportation vehicle such as busses, tour vehicles, taxis, cruise ships, private airplanes, commercial airliners, and others – you must have permission from the person in charge, captain, pilot, driver, ships communications officer, etc.

ARGUMENTS AND BRU - HA - HA'S

We Amateurs are heard not only by ourselves but many non-hams, shack visitors, folks on scanners, etc. We need to keep it from sounding like the CB band and keep a professional approach to our communications. Insults, jamming, disparaging comments with no ID, swearing, heated exchanges and calling another a LID (poor operator), really have no place on the repeaters. Good-natured jibing is self-evident and humor is in order, but grim enemies have been made on the repeaters -- bite ye olde tongue, keep cool or stop transmitting if things get heated. Don't stomp on new comers or any other comers for poor operating practice -- a gentle reminder or better yet - discuss it off the air.

And a word to Old Timers (OT's), don't lay the trip on the newcomers about "Why when I got licensed in 1901, why we had to blah blah, you young guys don't know ___, you have it easy blah, blah. Extra Lite. Why I had to take a 20 WPM code test blah blah and draw a circuit of a Colpitts Oscillator blah blah". That was then -- this is now. Make the new guys/gals feel welcome and not second class whatevers. We need to get new people in the Amateur Radio Service, or it will wither away.

Some of these newcomers can run circles around we old timers when it comes to computers, data modes, etc. Some of the best operators on our local repeaters are folks with less than a year as a Ham. Many have entered into emergency communications, PR events and NTS service. Be careful when talking to the newer callsigns, they may know a heck of lot more than you do in certain areas. But on the other side of the coin, Old Timers (OT's) pioneered Amateur Radio, made many important technical contributions (and still do), and we owe a debt of gratitude to those who kept Amateur Radio alive and well for almost 100 years. So a salute to those who served Amateur Radio so well and still do.

ABSOLUTE NO NO's

Do not engage in any business communications, your's or any else's. See FCC Rules, Part 97.
Do not allow any one to use your radios without your presence and supervision. See FCC Rules, Part 97.
Do not use foul language, innuendoes, off color language, dirty jokes, etc. See FCC Rules, Part 97.
Do not acknowledge Jammers in any manner. See Chapter 10.

Do not transmit on a repeater output frequency, very disruptive to repeater operation Use a simplex freq. Do not transmit on the repeater input without operating split. Very disruptive to repeater operation.

MULTIPLE CONVERSATIONS

Not all conversations are strictly two-way. Three, four or five or more Hams can be part of a "roundtable" conversation (five or more can be pretty unwieldy). A roundtable is a lot of fun, but it poses a problem; when the person transmitting is finished, who transmits next? Too often, the answer is everybody transmits next, and the result is a mess (doubling, tripling). You'll hear this often.

The solution is simple -- when you finish your transmission in the roundtable, specify who is to transmit next. "... Over to you, Fred." This takes some concentration as to who all is in the round table, try and keep track or write it down for reference. If you lose track -- turn it over to someone you know always keeps track -- "help me with the rotation Ted".

Also try "over to whoever is next" and hope for the best. Jump ball sez you don't know who is next-- hope for the best. Examples – "over to Fred, then Charlie". Or by implication, if it is Dave's turn - Dave are you still on I-15?"

Another style is 3 or 4 in the conversation where strict rotation isn't used, some one asks a question or makes a comment and another with the answer or counter-comment comes in after the beep, but careful not to double with someone else. Usually these are very short transmissions and comments -- sometimes called quick key QSO's. A technique to prevent doubling is to double clutch -- that is key up, immediately drop it and listen to see if no one else is transmitting, then go ahead.

LONG CONVERSATIONS

Repeaters are a great place to get information, SWR, Antennas, Radios, Computer stuff and just about anything else and some of these can get quite lengthy. Many clubs have several repeaters and when a long conversation is forming - especially during repeater busy times (commute time), maybe it is best to go to

a less used channel or to a simplex frequency. This is especially true if the conversation is of interest to just a few folks, such as butterfly collecting. So if the "prime" repeater is in use, try a secondary one.

PRIVATE CONVERSATIONS

Believe me there is no such thing on the VHF/UHF airways! If you change frequencies to a simplex frequency or another repeater for a private QSO, chances are someone will follow you and someone always seems to be listening. Scanners can find you very quickly. Even if you went to 1.2 GHz simplex, 2M SSB, or HF! Be careful with gossip on a "private" channel or frequency, it can come back to haunt you.

BREAKING IN

Repeaters are shared resources -- one big party-line. There are many times and reasons that a conversation in progress might be interrupted. You might break in to join the group and add your comments to the subject at hand. Someone else might break in to reach a party who is listening to the repeater. The breaker might have to report an emergency, traffic advisory, or make an urgent contact.

If you have a two-way conversation going, even somewhat of a private nature, expect that someone will break in. If this bothers you, choose a less often used repeater, or go simplex. Also the less used bands, 220, 440 or SSB portions will allow some degree of "privacy"

For breaking in - here are some suggestions: Pick a good time. If you have an emergency, a good time is NOW. That's why there's a pause between transmissions. Otherwise, listen a bit. Read the ebb and flow of the conversation. One of the fastest ways to establish a reputation as a jerk is to frequently butt your way onto the air without regard for the people already talking, and say -- just saying hello, or just letting you know I'm here. Give your call, and say what you want.

When you've listened and decided it's OK to break in, transmit before the beep and say something like this: "AC6V, I have an emergency," "AC6V, can I make a call?" or "AC6V joining in." "AC6V comment" "AC6V question" "AC6V info". But some common sense is in order, if a ham is giving directions to his/her out-of-state buddy - why would you break in and start a QSO on hiking?? Wait until they finish, of course.

What about "Break?" The problem with just plain "break" is that nobody knows exactly what it means, and everybody has to stop and find out. You will hear some Hams use "break" and meaning, "I just want to join in or make a call," "break-break" may mean "I have very important traffic," and "break-break-break" means, "I have a dire emergency." That's fine, but not everybody knows that. Plain English works much better. State the reason for your break, AC6V joining in, AC6V info, AC6V comment, AC6V traffic report, AC6V quick call please, AC6V emergency. If someone is joining in -- perhaps say, "callsign recognized, you follow Sally". However a "breaker" should be given the frequency immediately - could be an emergency. The ARRL discourages the use of break – see URL: http://www.arrl.org/tis/info/pdf/repeater1.pdf

If someone "breaks" in on you. What do you do? Let them transmit, RIGHT NOW unless you know absolutely and for certain that they do not have an emergency. Some break in just as you were ready to answer a question and answers it for you -- irritating as you may have spent some time and effort to look up the answer.

But you should immediately recognize the break as they may have an urgent call or real emergency. If folks would announce why they are breaking -- it would make repeater life a lot more organized.

Maybe somebody hasn't read this guide and isn't the expert operator you are now, and they just say "break" or drop in their call, when what they really mean is "HELP." So let them talk. Say "go ahead, or go ahead the breaker" And if they're one of those boneheads who's interrupting your perfectly good conversation for no reason but to hear themselves talk, well, bite your lip and be glad you know better.

The exception is when someone actually announces an emergency. Then CLEAR THE REPEATER! DO NOT TRANSMIT.

The station who declared the emergency has the frequency, and unless they ask for your help, don't give it. Unless... always and unless... they obviously don't know how to handle the situation... and you DO. (You do, don't you?)

ROLL CALL OR A DEMO PLEASE

When you have visitors to the shack and would like to demonstrate repeater FM and Amateur Radio, "AC6V could I have a demonstration please for some visitors. Callsign, name, location, and power." Hopefully not every body doubles -- try double clutching. Maybe give the visitors the brag tape on repeater range, mountain location, and the quality of Frequency Modulation.

SIGNAL REPORTS

It is important for we amateurs to maintain a high level of quality about our transmissions. In an emergency, you need a good solid signal into the repeater with adequate deviation and clear audio to expedite communications. If you are in a poor location or using an HT -- your signal may be weak and have path noise (repeater limiters not engaged) and you may drop in and out of the repeater so no one can copy.

A typical value for engaging the limiters might be about 0.5 to 0.6 microvolts, but this depends on how the repeater is set up. The repeater may pick up signals as weak as 0.2 microvolts again depending on how the repeater is set up. Raising power may help and a better antenna will certainly help.

When giving reports -- tell the transmitting station if they are full quieting and have adequate deviation and clear audio. Low level audio will make it difficult for mobiles with road noise to copy you. Bassy muffled audio is hard to copy -- cured with a better microphone. Describe the degree of path noise -- none, small, moderate, heavy or make a subjective percentage e.g. 90% quieting. Sometimes, low audio can be cured with being closer to the microphone or raising your voice, but if these don't help, an internal deviation control may have to be adjusted. If you are dropping in and out of the repeater with lots of path noise, DON'T shout into the microphone, this causes the available audio to spread (deviate) across a wider bandwidth and may cause excessive dropouts.

If you dropping in and out of the repeater, keep your transmissions short and have others assess whether or not you are getting into the repeater with an intelligible signal.

Giving signal reports of 59 (HF stuff) means nothing on FM repeaters as there is no way to tell what the signal amplitude is -- if it is a full quieting signal. If you report loud and clear - one can be loud and clear and still have path noise -- so it isn't a very good way to report signals on a repeater. Always keep in mind that the other station is coming through the repeater, you are not hearing them direct.

You are coming in loud and clear here in El Cajon means nothing -- it is the repeater that you are hearing not the other station direct. Receiving you 5 by 5 - (Military or Police style) doesn't mean much to the average repeater user either. Best describe the degree of quieting and the audio level and quality.

MAKING AN AUTOPATCH CALL.

Most repeaters have specific rules and require membership for using the autopatch. You must ask the trustee for the details. Learn the autopatch procedures and practice them often. Practice is best conducted during the wee hours, when usage and monitoring is at a minimum. For some reason, while the temptation to conduct business on most Amateur Radio modes is small, on autopatch it has been a big problem. This is probably because autopatch is the only mode that puts hams in contact with non-hams via Amateur Radio on a regular basis. It seems so easy to call work and pick up a message, tell the boss you're going to be late, etc. But remember, not just your business, but also the business of the party you call. And that goes for non-profit businesses and government agencies as well! The FCC has been very clear about this.

If you try to make a call on an autopatch that even hints of being business-related, you can expect a repeater control operator to terminate your patch. Following are some generic procedures just to get you started – please do not try and use these. You must have the formal official guidelines from your repeater trustee or designated autopatch guru and use their procedures. Long distance calls are usually not allowed.

On some repeaters, the autopatch allows for dialing only pre-programmed personal and emergency numbers. Some limit calls to three minutes, except for emergency numbers. Autopatchs usually require an access code sent in Touch-Tones. However, some repeaters have open autopatch, allowing anyone to use it. These repeaters usually use an access code of "*". Other repeaters have "closed autopatch" where the patch codes are disclosed to members only. The procedure for using an autopatch depends on the policy of the repeater's sponsor, and on the design of the control system.

Here is a generic guide. Always make sure the repeater isn't busy. If you've just turned on your radio and don't want to wait to see if the repeater is active, you might announce, "AC6V is this frequency in use?" If you have a clear repeater frequency, then identify "AC6V accessing the Autopatch", dial the access code, unkey your transmitter and listen for a dial tone. Then dial the number. Some repeaters will mute the touch-tones.

On other more sophisticated repeaters it would go like this. Dial the access code and telephone number as one string. The controller stores and then regenerates the number. You then hear it pick up the line, send a dial tone, dial the number, and ring. The patch control holds the repeater transmitter on, with the phone audio connected. When you transmit, your audio goes on the phone line. The party you called answers, and you transmit and talk.

Most new people on the telephone end get confused by autopatches until they've had some experience. When you're transmitting, they can't interrupt you, but they may not be aware of this. You can reduce this problem by keeping your comments short, and releasing your transmit button after your last word.

Tell the party on the telephone that you are talking to them via amateur radio phone patch and it is one-way (not like a cell phone) – I talk, you listen, and vice versa. Using over is helpful. Most "civilians" have enough movie experience to know how it works. When you're finished with the call, say good-bye and have your party on the other end hang up.

Then send the disconnect command which on many autopatches is the "#" key. And identify again, "AC6V clear autopatch". Listen to make sure you successfully killed the patch (the repeater may talk to you, or beep, or just drop).

NETS

A net is a group of several stations that meet on a specified frequency at a scheduled time. The net is organized and directed by a net control station, who calls the net to order, recognizes stations entering and leaving the net, and authorizes stations to transmit. Most nets are "directed nets" so as to keep order amongst the multitude tuned in. You should not transmit unless solicited or as directed by the net control. Examples - "check-ins please". "AC6V -- do you have anything for the net." If you are not part of the net - use a different repeater for other matters.

Nets can be of a wide variety of subjects: NTS (National Traffic System), Ham Help, Morse Code Practice, Offroad Nets, Language nets, Youth Groups, Sailors/Boaters, Emergency - ARES, RACES, Microwave, ATV, Facetious Group, Field Day Planning, Swap, Trivia, YL, and many others.

Swap nets are held on some repeaters - yes it is legal to announce equipment for sale or items wanted on Amateur Radio as long as the items are Ham related. Since we use computers for Ham Radio, these have crept into the sell/wanted lists.

EMERGENCY NETS

Emergency nets are an organized meeting of Amateurs to conduct dissemination of emergency news, drills or the real thing. Most organizations require some training on how to conduct and handle emergency traffic. Emergency groups include Amateur Radio Emergency Service (ARES) and Radio Amateur Civil Emergency Service (RACES). Contact your local emergency group or ask on a repeater for how to join. – see URL: http://ac6v.com/emergency.htm for more details on emergency service. Joining these groups is a good way to get to know people as well as be prepared to help in a real emergency. These groups also participate in Public Relation events, such as parades, air shows, horse races, street fairs, etc. Participating can be a great way to learn how to pass traffic and understand communication protocol.

HAM HELP NETS

Many repeaters sponsor ham help nets for new comers or anyone else who has questions or needs advice. Subjects often discussed are antenna building, SWR, charging batteries, computer problems, HF operation, microwave, ATV, PSK31, the list is endless. Some conduct Morse code practice sessions.

One can get on the repeater at any time and seek help, but there may be a limited audience. Ham help nets usually draw several experienced hams who can address just about any ham related question. The moderator typically has the Internet up and running and can direct questions to applicable URLs.

TECHNICAL NETS

A great way to get into the more specialized areas of ham radio is to join in a technical net. These include but not limited to: Antennas, Microwave, ATV, Field Day, Computers, Backpacking, Offroad, and others.

FIELD DAY AND CONTESTS

The ARRL sponsors a field day -- Always The Fourth Full Weekend In June. Here is another way to meet people and get on VHF/UHF and HF and experience the specialty modes, ATV,PSK31, RTTY,etc. Field day now includes a Get-On-The-Air (GOTA) activity which allows Technicians and the general public to operate on the contest bands under the supervision of a control operator with the appropriate privileges. URL: http://www.arrl.org/contests/announcements/fd/

QUICK TROUBLESHOOTING

Here are some tips when your radio isn't working properly.

Can't access a repeater:

1. The transmit frequency may not be correct, check offset frequency and duplex plus or minus.
2. On some rigs, the split function must be activated.
3. For Pled Repeaters, the tone frequency must be correct and the tone function activated. A "T" indication is typical when the tone is activated.
4. You are too far away from the repeater with your current antenna and power. Try raising power and use a better antenna.
5. You are 5 kHz off the proper frequency. Typical of older units where the 5kHz was a separate button.
6. The repeater may be off the air (work party).
7. Repeater is usually non-Pled, but a jammer or stray carrier occurred, so repeater was put on PL. Activate the proper PL tone.

Radio won't transmit.

1. The S-Meter or "bars" should indicate power out and if in repeater operation, the frequency readout should change from the receive frequency to the transmit frequency as you press PTT.
2. Be sure the VSWR is not too high, i.e., below 2.5 to 1.
3. Check that you are not trying to transmit out of the prescribed Amateur band. Maybe wrong plus or minus split.
4. Check that the batteries aren't low, most rigs will alert you of this. Some radios will receive fine but not transmit at all on low batteries.
5. Some PTT switches take considerable pressure to engage, press firmly.

CHAPTER 8: PHONETICS, CALLSIGNS, Q-SIGNALS

PHONETICS

On the HF bands, 1.8 to 30 MHz, fading, interference, and weak signcals can make copy difficult at times, so for many years Hams have used an internationally accepted method of Phonetics for the letters of the alphabet. Alpha, Bravo, etc., sometimes called the NATO/ITU Phonetic Alphabet or Military Phonetics. They are also used on VHF/UHF as these help differentiate between similar sounding letters. Note that the numbers are Military recommendations.

A - Alfa	G - Golf	M - Mike	S - Sierra	Y - Yankee	4 - Fower
B - Bravo	H - Hotel	N - November	T - Tango	Z - Zulu	5 - Fife
C - Charlie	I - India	O - Oscar	U - Uniform	Ø - Zay-Roh	6 - Six
D - Delta	J - Juliet	P - Papa	V - Victor	1 - Wun	7 - Seven
E - Echo	K - Kilo	Q - Quebec	W - Whiskey	2 - Too	8 - Ait
F - Foxtrot	L - Lima	R - Romeo	X - X-Ray	3 - Tree	9 - Niner

Some folks use DXing phonetics -- such as King Six Japan Norway instead of Kilo Six Juliet November. This is common on HF but not on VHF/UHF. Others use police phonetics, King Six Adam Mary, but both DX and police phonetics best not be used on VHF/UHF as many are not familiar with them. And still others use cutie phonetics, Karl Six Always Killing Time; these make a person's callsign easy to remember and are often used. Names can be cuties as well -- Rod becomes "Ragged Old Dog" or Bob is "Broken Old Bottle". Probably preferable to use the NATO phonetics, but the cuties do make it colorful and memorable and you will hear them often. Perhaps use NATO phonetics when you first talk to a new contact and if they can't remember your call, use a funny phonetic. AC6V = Alternating Current Six Volts! Using NATO phonetics will prepare you for emergency service and the HF bands.

For other slices of Ham Speak -- see Origins and Ham Speak. And Jargon. http://ac6v.com/73.htm. If you are new to Amateur Radio -- you may be interested in the History of Ham Radio. http://ac6v.com/history.htm

Q-SIGNALS

Q- Signals were born in the early days of wireless radio while using Morse code. They are a short hand method of sending common messages. CW maritime and the military used upwards of 200 Q-Signals for Aviation, Marine, Weather, and other uses. Also a set of QN signals have been developed for net use. Expert Ham CW ops use upwards of 50 Q signals, but only 9 of them are common on VHF/UHF FM repeaters. Other Q-Signals are discouraged only because many repeater users have no idea what they mean. QTR -- What ?? (Just ask for the time!) Use of 10 codes are also discouraged for the same reason. What's your 10-84 --- huh? (Just ask for their phone number!). And 10 codes smack of CB lingo.

When voice modes became available, Hams on HF carried Q-Signals over to AM, then SSB. With the advent of VHF/UHF FM, some have carried over to FM Repeaters. The Q-signals used on HF are also used on VHF/UHF SSB and CW. They are really not necessary on repeaters, but many hams use them so best know what they are.

AC6V's Guide To Ham Q-Signals Commonly Heard On VHF/UHF Repeaters

Q-Signal	Question ?	Answer, Advise, or Order
QRM	Are you being interfered with? Do you have interference	I am being interfered with. *Common Usage - A jammer or a spurious signal is interfering with the repeater.*
QRN	Are you troubled by static?	I am troubled by static. *Common usage "Path Noise"*
QRP	Shall I decrease power?	Decrease power *(commonly a Low Power Station)*
QRT	Shall I stop sending?	Stop sending *(Common usage - clear of the repeater) (Maybe Shutting Down*)
QRX	When will you call me again?	I will call you again at ... (hours) on ... kHz (or MHz) *Common Usage Pause, Wait, Standby, maybe a phone call or shifting gears in a mobile*
QRZ	*Who is calling me?*	You are being called by ... on ... kHz (or MHz)
QSL	Can you acknowledge receipt? *Common usage - Did you get that OK?*	I am acknowledging receipt. *Common usage -- Yes I understand, Roger*
QSO	Can you communicate with ... direct or by relay?	I can communicate with ... direct (or by relay through ...) *Common Usage -- A 2-Way Contact, a conversation*.
QSY	Shall I change to transmission on another frequency?	Change to transmission on another frequency (or on ... kHz (or MHz).*Common Usage - Changing channels (frequencies*)
QTH	Common usage *Where are you located?*	*Common usage - Your location mobile or home*
73	73 is not a Q- It means "Best Regards". Since regards is plural we say 73 not 73's.	
88	88 is not a Q-Signal but is a hang over from telegraph days. It means "Love and Kisses"	Commonly used for exchange with the opposite gender, but used for those who you would normally send " Love and Kisses" in a correspondence.

Note that on VHF/UHF SSB & CW you are more likely to hear the same Q-Signals as on the HF Bands.

If you want to see the list for SSB/CW use on HF/VHF/UHF see URL: http://ac6v.com/Qsignals.htm

An all time Mega Q-Signal list is at URL: http://www.kloth.net/radio/qcodes.php

QN Signals can be found at URL: http://home.alltel.net/johnshan/cw_ss_list_qn.html

---

CALLSIGNS

There was a time when you could tell where a Ham was residing by their district number. But this was changed to allow Hams to keep their callsign when they relocated. In the USA, the time is long past that you can tell where a USA ham resides. You can have a W2 call (NY & NJ) and live in California (W6). For USA hams, it is not necessary or required by the FCC to use a portable designator. Often visiting stations will ID portable e.g., W4XYZ - Portable W6. This is not required under FCC rules, but is a way of the visiting station letting others know they are from out of state. For USA Districts and states -- see USA Map. http://www.arrl.org/awards/was/map.gif USA

USA calls including islands and possessions are WA-WZ, KA-KZ, NA-NZ, but only AA-AL. AM belongs to Spain, etc -- See Prefixes. http://ac6v.com/prefixes.htm

For hams from outside the USA operating on reciprocal licensing agreement, CEPT license, or IARP permit, they ID by the US call district identifier, followed by their non-US call sign. Example, W6/G1ABC. They might say G1ABC portable W6. This ham is operating in California and is licensed in the UK. See World Wide Prefixes. For USA Hams operating overseas -- See Licensing Abroad (www.ac6v.com)

Continental USA--WA-WZ United States (XX)= State Abbreviation

WØ (CO) Colorado	W3 D.C. ITU	W6 (CA) California
WØ (IA) Iowa	W3 (DE) Delaware	W7 (AZ) Arizona
WØ (KS) Kansas	W3 (MD) Maryland	W7 (ID) Idaho
WØ (MN) Minnesota	W3 (PA) Pennsylvania	W7 (MT) Montana
WØ (MO) Missouri	W4 (AL) Alabama	W7 (NV) Nevada
WØ (NE) Nebraska	W4 (FL) Florida	W7 (OR) Oregon
WØ (ND) North Dakota	W4 (GA) Georgia	W7 (UT) Utah
WØ (SD) South Dakota	W4 (KY) Kentucky	W7 (WA) Washington
W1 (CT) Connecticut	W4 (NC) North Carolina	W7 (WY) Wyoming
W1 (ME) Maine	W4 (SC) South Carolina	W8 (MI) Michigan
W1 (MA) Massachusetts	W4 (TN) Tennessee	W8 (OH) Ohio
W1 (NH) New Hampshire	W4 (VA) Virginia	W8 (WV) West Virginia
W1 (RI) Rhode Island	W5 (AR) Arkansas	W9 (IL) Illinois
W1 (VT) Vermont	W5 (LA) Louisiana	W9 (IN) Indiana
W2 (NJ) New Jersey	W5 (MS) Mississippi	W9 (WI) Wisconsin
W2 (NY) New York	W5 (NM) New Mexico	
	W5 (OK) Oklahoma	
	W5 (TX) Texas	

Alaska uses a 7L such as KL7xyz, NL7xyz. Hawaii has an H6 such as KH6xyz, NH6xyz. And US islands use an H also such as KH2xyz for Guam. See Prefixes at URL: http://ac6v.com/prefixes.htm

USA callsigns are kinda hashed up as to operator class, but in general, callsigns may reflect the license class, although many keep a lesser callsign when upgrading. Example, KM6SN is an extra but chose to keep an original Advanced call.

A 2-by-3 [two letters, a number, and three letters] callsign typically belongs to a Novice or Technician operator, e.g., KC6UQH. Also some of the newer General operators have a 2-by-3 call.

A 1-by-3 callsign usually belongs to older Technician and General class operators. e.g., N6AEK. These callsigns were 'used up' in the sequential callsign system, so newer Tech and General operators are now receiving 2-by-3 calls such as KG6DVD.

A 2-by-2 call usually belongs to an Advanced or a newer Extra class operator. An Extra 2-by-2 usually starts with an 'A'. e.g., AC6HZ. An Advanced 2-by-2 usually starts with a 'K', 'N', or 'W'. e.g., KB5MU

A 1-by-2 or 2-by-1 call usually belongs to an older Extra class operator. e.g., N6ST, AC6V. When these sequential callsigns were 'used up', the Extra class operators started receiving 2-by-2 calls starting with the AA block. However newer extras can apply for 1-by 2 or 2-by 1 vanity call if available.

A 1-by-1 callsign is very special. They are assigned to special event stations only for a short period of time, e.g., WØW. They are not assigned to individual operators as their regular callsign

The vanity callsign program allows folks to get a new or old callsign. For example KD6JXY got K6JXY without another exam. But vanity calls are only available in your license class or below, i.e., KG6DVD can get K6DVD (if available), but not K6DV.

REPEATERS THROUGHOUT HISTORY

522 BC Persian Army employs a relay system where soldiers positioned on hilltops shout and relay military messages 30 times faster than by runner. Accounts of flags, mirrors and smoke signals appear in early history.

1876. Humor Time - Battle of the Little Big Horn – Custer leaves the HTs, Gatling Guns, and Sabers back at Fort Abraham Lincoln. His opponents use smoke signals and know Custer's every move.

1941. As the USA enters WWII, Ham Radio Operation is suspended except WERS. War Emergency Radio Service (WERS) was created in June. 1942. WERS operated on the former amateur 2 1/2 meter band (112-116 MHz).

1945. Hams were given authorization to begin operating again on the 2 1/2 meter band, on a shared basis with WERS. WERS was terminated in mid-November. By the 15th of that month, the FCC released bands at 10, 5, and 2 meters for amateur use.

Mid to Late 1960's Amateurs start to build 6M AM and FM repeaters and by the mid 1970's, many repeaters are in operation. Equipment used vacuum tubes and crystal pairs for frequency control. Some AM repeaters were built. Heathkit had 2ers and 6ers that many hams built. Hams grind surplus crystals by hand to the desired frequency. Some early repeaters were separated geographically to avoid receiver desense as duplexers were not common.

1972. A national band plan is announced for 2 meter FM, the national simplex frequency is established at 146.520 and the FCC released the first repeater rules. Logging requirements are relaxed.

1972 Sept 9th, - The Palomar Amateur Radio Club in San Diego, CA received the coordination for their 146.730 all vacuum tube repeater on Palomar Mountain from the newly established Southern California repeater coordination body in Los Angeles at their first conference. However the club had been successfully operating a test repeater in a garage in Vista during 1971. The duplexer was made from discarded shell casings obtained from a Navy Battleship! Subsequently the entire system was stolen, hence locked repeater shacks.

1974 - WR prefixes began to appear on repeater call signs. **1978 -** WR repeater call signs are phased out.

1977 repeater rules were simplified further. A new repeater sub-band is established at 144.5-145.5

CHAPTER 9: FUNNY REPEATER SOUNDS, MYTHS

SOME FUNNY SOUNDS ON THE REPEATER

As you listen in on repeaters some strange stuff can be heard.

Capturing. If two stations are transmitting at the same time (doubling) on the same frequency – it creates problems. If the two stations are about equal in strength to the repeater, both may be heard, typically with a heterodyne or beat note. When one of the stations is stronger than the other, the stronger station "captures" the weaker one and the strong guy wins. This is in contrast to AM or SSB where both may be heard. So if you have a weak signal and double with a strong station, you will not be heard!

Picket Fencing. This is caused by a station in motion alternately and quickly entering a good then bad transmitting location. This sounds like someone rubbing a stick across fence boards. Typically, caused by a mobile in an underpass or bridge crossing.

Morse Code. Repeater transmits something in Morse code - many repeaters ID in Morse or you may hear something like dit dah dah dit (P), this can mean several things depending on the repeater configuration -- it has switched to emergency power or has reset the controller. Some send an alarm if they are monitoring a function -- one in Northern California sends H2O when the water tank it guards is at a low water level. Ask the repeater trustee for meanings. Locally one repeater sends (P) when the microprocessor has been reset -- maybe a power glitch, but maybe not, a gadget called the "watchdog timer" can detect anomalies and reset the microprocessor to its boot value. It then has to be signaled by a control operator to return to normal mode -- usually by a link - perhaps 440 MHz or a landline.

Station Doesn't Hear You. A station (usually weak) solicits a QSO or radio check -- but doesn't hear you when you reply. Perhaps they are on another repeater that is on the same repeater pair and don't hear you.

Can't Access A Repeater. You are receiving a repeater very strong -- but can't access it. Probably your offset or PL is not set correctly. Or the repeater is an alligator – big mouth, small ears.

Station is Calling, But No Beep. You hear a transmission -- but there is no beep or hang time -- the transmitting station is probably transmitting on the output frequency (does not have their radio set for offset transmission). They are operating simplex on the repeater output. Easy to determine if the offending station is mobile – your S-Meter will bobble around. Talking on the output of repeater is confusing and disruptive to repeater listeners and the repeater may come on over them. A definite no no!

Station Doesn't Answer. You hear a transmission – soliciting a QSO, there is a beep and squelch tail, but they don't respond to you. Maybe their volume control is turned down. Best not to continue since they can't hear you.

Distortion. A station comes on with distortion -- more times than not this is caused by them being off frequency -- usually 5 kHz. Tell them to check their frequency setting and quote the repeater output frequency down to the last digit, e.g., 146.730 MHz. If the signal is extremely loud and distorted -- could be over deviation. Or could be a "broken radio".

Rattle, grumph, bloop. This could be intermodulation interference. Happens a lot to HTs with a high gain mobile antenna in a high RF environment area (Pagers, Utilities, etc). Intermod is false or spurious signals produced by two or more signals mixing in a receiver or repeater station. Or intermodulation distortion (IMD) -- the unwanted mixing of two strong RF signals that causes a signal to be transmitted on an unintended frequency.

Path Noise - Noise on your signal. For the most part FM is noise free, however your signal must be strong enough to engage the limiters in the repeater receiver. The static or noise is called "path noise" and is due to a variety of causes, which amounts to amplitude variations on the signal picked up by or in the repeater receiver. Your signal is too weak into the repeater. Since the repeater limiters cannot "clip" the AM on weak signals, the repeater "repeats" the noise on the transmitter output. Try increasing power, a better antenna, or with an HT walk around to where you receive the repeater the strongest and try that location as your transmit point.

White noise is a scientific term used to describe a spectrum of broadband noise but is often incorrectly used in place of path noise. White noise is a type of noise that is produced by combining sounds of all different frequencies together. If you took all of the imaginable tones that a human can hear and combined them together, you would have white noise.

Trivia -- Pink Noise -- unlike white noise, pink noise bears a logarithmic characteristic, and as such, represents a psychoacoustic equivalent of white noise sweetened for human ears. This is the signal used to test speakers and set equalization in theaters and other venues. When you tune up your home multimedia system, the noise used to drive the speakers for the volume settings is probably pink noise.

ZZZZZ Impulse Noise. In your mobile, ignition noise can sometimes be so severe as to be heard in your receiver but not transmitted to the repeater. This can be solved with proper shielding. See Noise Suppression Techniques. http://ac6v.com/opmodes.htm#MO

Whine. Sometimes you will hear a whine on a mobile signal that varies in pitch as the engine RPM increases. This is due to the auto's alternator AC output reaching the radio's circuits. This can be filtered out with coaxial capacitors or chokes. See Noise Suppression http://ac6v.com/opmodes.htm#MO

Motor Boating. Radios can go goofy and introduce feedback at the wrong time and the wrong place and the result is a put put sound similar to an outboard motor sound. Said radio is broke.

Buz Buzz Hum Hum. The transmitting station is operating with an AC Power Supply and not all power supplies are created equal and may have insufficient filtering and AC will appear on the transmitted signal as AC Hum. Solution --- get a better power supply or make a filter for the one with the problem. Most HT battery chargers will NOT support transmitting and this is a common problem when you hear the hum. Keep in mind that a 50-Watt Radio can draw up to 11 Amperes on transmit, so an adequately filtered power supply is a must when operating as a base station. Another Buzz Buzz is inadvertent Packet transmissions on a repeater.

Whoosh Woosh. Wind noise from open cockpit mobiles, boats, or mountain climbers. There are noise canceling mikes and accessory wind screens for this one. Placing foam rubber over the mic element may help. Some motorcycle mobiles use this fix.

Kerchunk -- Characteristic noise made when the repeater is brought up -- no one says anything and it drops -- usually pretty fast. Someone is testing to see if they are accessing the repeater. This is illegal for one thing, annoying for another. Better just say "AC6V Testing, No Response Necessary". Some repeaters have a "kerchunk filter" to reduce key-ups from stray signals.

Desense. You and another mobile are very close to each other and going through a repeater when all of a sudden -- you can't hear one another. Well your transmitter is overloading the other person's receiver in this situation. Try simplex or low power or increase your distance to each other.

Over Deviation -- extremely loud audio with distortion on the audio peaks. Caused by improper internal adjustment and/or shouting into the microphone. Said radio or operator needs adjustment. Internally this is called a Dev adjustment. Not for the faint of heart on the micro-miniature component radios.

Under Deviation -- symptomatic of soft-spoken persons or speaking too far away from the mic. Everyone strains to hear them with volume controls wide open and mobiles with wind noise give up trying. When the next station comes in -- the wide open volume control blows listeners out of the room or car. Oouuch my ears!! Said radio needs a dev adjustment or a conscious effort of the operator to speak up (close-mouth the mic).

Shush glep shush glep -- The calling station has a pumping sound. Their battery is going dead.

Breathiness on the Audio. Try cross-talking the mic at an angle rather than straight into the microphone.

REPEATER AWARDS (Just for fun)

TOR Award – Timed Out Repeater Award. Awarded the first time a user times out a repeater. Most of us achieve this prestigious award at least once.

TORA -- This is awarded to the user who has timed out the repeater 10 or more times. Never pauses to reset the timer. Speaks in paragraphs and chapters.

Quick Key Karl - Never waits for the beep but usually does not time out the repeater.

Masked Person Award - Never Identifies.

Cowardly Lion -- Throws in disparaging remarks without IDing. Usually a graduate of the CB Band.

Macho Man - Uses swear words to re-enforce their macho-ism.

HF Harry -- Uses a multitude of Q Signals and HF jargon. "QRZ I'm QRU, and QRV on the frequency and tuning - Willy Dog Six Dirty Underwear". Huh Whatta they say and why are they tuning?

Orator Ollie - Goes on and on -- Maybe Practicing for Toastmasters. Has earned the TORA award with numerous endorsements.

Drop in Dave - Jumps into every QSO - even if she is telling he -- to get some lettuce.

Lecturing Larry - Constantly lectures newcomers and old timers on protocol. Usually has a cop badge. Hey if you or others have told old Abner a hundred times to wait for the beep and he ain't getting it -- maybe just give up and send him an e-mail.

Technical Terry - Instructs disinterested listeners on the technical aspects of Ham radio. Lots of non-tech folks on Amateur Radio these days -- so judge your audience. Answer if they ask, but no need to educate the world if they could care less.

CB Charlie - "Hey KAA9QZY, you got your ears on. Good Buddy?" You're hitting me with 10 pounds. There is a City Kitty up ahead - better slide that Dry Box to the Granny Lane 10/4. I'll pass the good numbers to ya. Come on back. What's ur 10-84. -- Drives old time hams into over deviation!

Callsign Carla - gives their callsign at the start and end of each transmission and in between, over and over again, even in a non-round table. -- See identifying requirements.

Power Mic Paul - Uses an amplified mic (misadjusted) or speech processor. Picks up the cat sneaking across the room and every sound for 30 feet! Sounds like they are in an echo chamber.

Loud Mouth Lester -- well Lester is Lester -- on and off the radio.

Pausing Pete -- when it is his turn - leaves 30 seconds delay after the beep, where did he go? Alien abduction???

REPEATER MYTHS

Often heard is "I'll raise power so I'll be louder". Since repeaters are frequency modulation not AM or SSB, as long as your signal has engaged the repeater limiters (full quieting), an increase in power will not make you louder, since the audio is proportional to the FM deviation (swing) and not amplitude. Raising power may help get rid of path noise thus making your signal easier to copy. Another is "I'm near the repeater, so that's why I am so loud". Nope – if you are full quieting – distance doesn't affect volume.

"You are putting out a great signal way down here in my location." In truth the transmitting station may be running a milliwatt but is close to the repeater -- and it is the repeater that has a gangbuster signal to you -- being it is 5600 feet up on a mountain and spitting out 20 Watts downward. If you want to hear the transmitting station direct – press the REV button (or equivalent) on your radio. This will only work for stations near you. Another is "You are 20 dB over S-9 (or your hitting all the bars on my radio)" -- NOPE -- the repeater is doing that, not the transmitting station. Tis true when you are operating simplex.

"WOW my HT is a super radio --- I'm getting into the repeater from way down here on 1 Watt and a rubber duck antenna!!" Modern HT's are indeed a wonder, but a rubber duck is a pretty lousy antenna -- designed for convenience, not performance. In truth the big repeater antenna way up in the sky and its hi-tech receiver is doing most of the work. "What is your location (QTH)?" Nice to know for chatting, but has nothing to do with the signal strengths, since the repeater is in between you and the other Ham.

CHAPTER 10: COPS AND JAMMERS

REPEATER COPS

It is best not to play repeater cop and criticize folks on the air about their protocol or use or overuse of the repeater. Leave this function to the control operator's discretion. Playing cop discourages beginners and some folks are very sensitive to on-the-air criticizism. If they ask about the proper protocol, answer them of course, but best discuss the matter on a landline, e-mail, club meeting or on a simplex channel. Or direct them to a publication or web site for repeater usage.

You can alert newcomers about observance of the beep and leaving a gap for others to get in, as this is critical in an emergency situation. Also some tend to time out the repeater over and over again by talking too long and not even aware of the timer function -- so gently inform them. If you know them well enough - maybe give them the TORA award -- Timed Out Repeater Again. But if an experienced Ham is not observing the beep or timing out the repeater over and over again, just give it up, it is useless and tiresome to keep reminding them.

DEALING WITH INTERFERENCE -- THIS IS VERY IMPORTANT -- PLEASE READ

If a jammer appears or some one not identifying, swearing, throwing carriers, etc --- DON'T acknowledge them, they are looking for attention -- don't give it to them. IGNORE IGNORE is the drill. It doesn't happen often, but it does happen -- it's a big world out there, and there are some nuts. Some of them find a Ham Radio now and then, and discover the delight of offending an audience and the power of holding down the PTT key. The key word here is audience, they are seeking to disrupt and get into a bru-ha ha.

Deliberate interference and bad language are designed to make you react. The person doing it wants to hear you get mad. They love it. And if they don't get it, they go away, usually quickly. So when you hear the rare nasty stuff on the repeater, our advice is to please ignore it completely. Don't mention it at all on the air. Sometimes a repeater control operator will decide that the best way to handle the situation is to turn off the repeater for a while, but the rest of us should bite our tongues and not mention the interference.

For the station that inadvertently sits on the mic and sends out a carrier, use a different approach. Listen carefully and sometimes you can hear faint audio which will ID the sitter. Look them up in the Club Roster and give em a landline. Otherwise, wait until the carrier drops and just announce "AC6V, open mic alert, someone has a stuck PTT, I'm sure it is inadvertent, but all stations check your mics please". Many times the open mic is a long one and will time out the repeater. Some machines won't come back on until the carrier goes away.

Hi Rod,

Well, I believe the following definitions to be accurate but as you know, on the Internet, you'll always get an argument:

Slim - Someone pretending to be a DX station, usually rare, that is supposed to be on the air. For example, someone in southern Argentina pretending to be Heard Island, VK0IR

Pirate - Someone using an existing callsign and operating on the air, e.g. claiming to be WA6YOO/4 on an North Carolina island group. What would happen is that I would suddenly be surprised by receiving a number of QSL cards and wonder "What Happened?"

Bootlegger - Someone, usually not a Ham but a wannabe, making up a callsign, one usually not in the callbook, and getting on the air. Sometimes it is someone who already bought a radio, took the test and flunked, and then gets on the air anyway.

These definitions are somewhat flexible because some of their aspects might be combined.

73 W6YOO

Origin Of Slim

The origin of the term SLIM is alleged as follows: Sometime in the mid-1960s, an operator showed up on 20m CW, identifying himself as 8X8A, claiming he was located on "Cray Island," an island that had just emerged from the ocean floor following a major undersea volcanic event. The operator said his name was "Slim." Huge pileups endured for a few days, then "Slim" was exposed as a pirate and he disappeared from the bands. Ever since, Pirates have been called "Slim." What made "Slim's" claim even partly credible in the first place was the emergence of Surtsey, an island near Iceland in about 1964. Surtsey rose from the floor of the Atlantic as a result of undersea volcanic activity. It was a major news event at the time.

CHAPTER 11: INSIDE A REPEATER

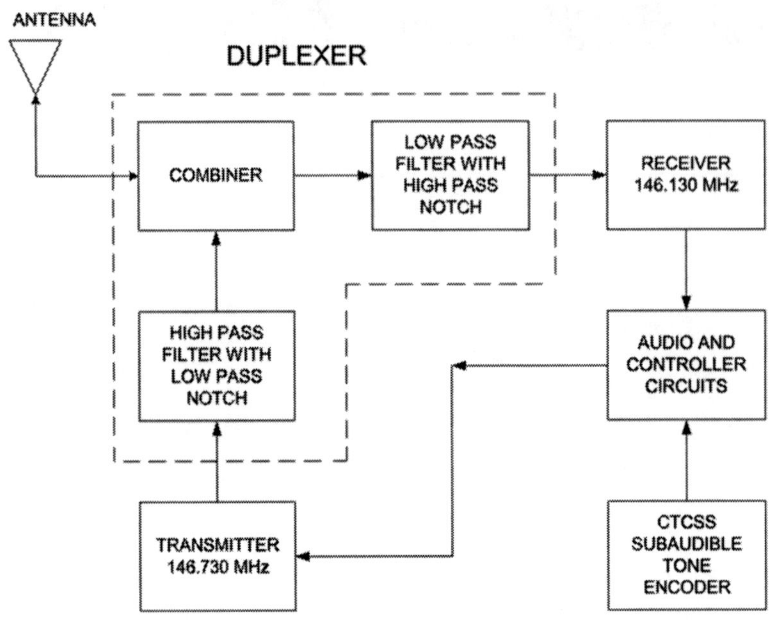

FIGURE 1. TYPICAL 2M FM REPEATER

Following is a very simplified discussion of a basic FM Repeater. The discussion is written from the perspective of a user rather than any in-depth technical analysis.

Repeaters can operate off of two antennas or just one antenna. In the latter, a repeater is able to receive and transmit simultaneously on a common antenna. So in our example in Figure 1 – a 146.130 MHz signal is being received and a 146.730 MHz signal is simultaneously being transmitted.

A key factor in keeping the transmit energy out of the receiver is a duplexer which allows the repeater to operate on a common antenna and transmit and receive simultaneously on fairly close frequencies. Several types are in use, but a typically one is a notch duplexer. Notch duplexers are large cavity type filters with very high Q and narrow bandwidth to the two frequencies involved and use tunable notches. The transmitter side of the duplexer passes the transmitter frequency but has a notch at the receiver frequency. The receiver side of the duplexer passes the receive frequency but has a notch at the transmitter frequency. This keeps transmitter noise out of the receiver and prevents the transmitter from overloading (desensing) the receiver circuits. The receiver amplifies the signal, clips off any amplitude modulation (noise) and detects the FM modulation as well as any PL, DCS codes, DTMF codes, etc. The audio components are sent to the controller for processing.

Repeater receivers can be set for very high sensitivity. Signals of a few tenths of a microvolt can activate a repeater (typical 0.2 uV). The limiters typically engage above that level (typical 0.6uV), so very weak signals may have path noise which sounds like "static" in the audio. For the most part FM is noise free, however your signal must be strong enough to engage the limiters in the repeater receiver.

The static or noise is called "path noise" and is due to a variety of causes which amounts to amplitude variations on the signal picked up by or in the repeater receiver. Since the repeater limiters cannot "clip" the AM on weak signals, the repeater "repeats" the noise on the transmitter output. Reports are : You have path noise.

For non-PL repeaters, a carrier is usually all that is needed to bring up the repeater. However, some non-PL repeaters may require audio as well. If the repeater is set on PL (CTCSS) and the transmitted signal contains the proper PL frequency, then the controller enables the transmitter. Some repeaters transmit a PL code even if it is not set for PL access. This was discussed in Chapter 4. This allows one to set their receiver to only open up if the repeater is sending the proper PL, termed tone squelch or tone decode. Also transmitted PL is required by some autopatch systems.

The transmitter amplifier boosts the signal up into the Watt level. A fairly high level machine here in San Diego runs about 20 Watts output. Gain antennas can be used for a higher effective radiated power. Some antennas have a down tilt pattern to radiate energy down from the mountaintops to the lower altitudes below. Repeater locations on the side of a mountain rather than on top can give a directional pattern. This is typical in large metropolitan areas to avoid interference when repeaters are on the same frequency.

Either another band or telephone lines are used to command the repeater from a remote location. The operator empowered to command the repeater is called a control operator. DTMF codes can be used to perform various repeater functions as well. The FCC has rules that the repeater must be capable of being controlled from another band or method such as a telephone line. He/she can shut down the repeater, place it or release it from PL, reset the system, and possibly monitor repeater status items such as battery power or commercial AC line power levels. Some repeaters have modes for normal, nets, and emergency. The controller performs a variety of other functions such as autopatch interface and more, DTMF codes can be used to perform various repeater functions as well. Fig 2 shows more detail of repeater workings.

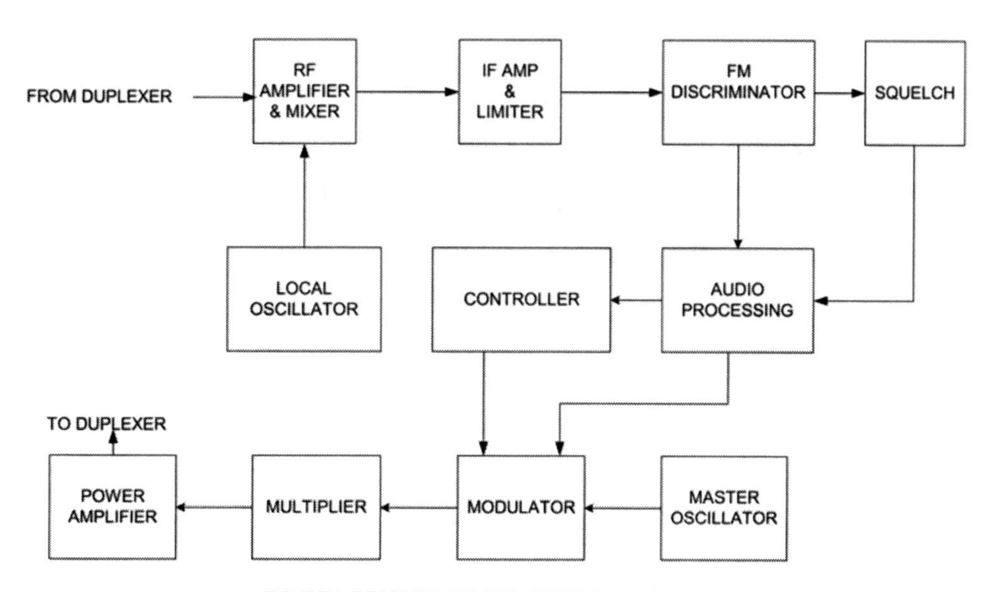

FIGURE 2. RECEIVER AND TRANSMITTER CIRCUITS

In the receiver, incoming signals are usually in the microvolt range and an amplfier is necessary to get the signal level up in level to process properly. The local oscillator mixes a signal with the RF signal to obtain an intermediate frequency. Using an intermediate frequency (IF) makes it much easier to design filters for selectivity, and overall this gives the advantages of a superheterodyne receiver. The advantages being good image rejection and selectivity. The limiters clip off amplitude modulation such as noise, impulse noise and any variations in amplitude. However the signal must be strong enough to engage the limiters otherwise path noise may be present on the repeated signal. The receiver circuits in your radio are similar to the repeater receiver circuits.

Next the FM modulation is detected and these are referred to as discriminators. FM discriminators can be of several types, Foster-Seely, Ratio Detector, or phase locked loop. Today's FM demodulators are contained within integrated circuits and the only requirement is for a coil and capacitor to be connected to the chip to provide the frequency dependent circuit. The discriminator is tuned to the carrier frequency so with no modulation the discriminator output is zero.

For example: below is an FM signal with no modulation – just the carrier – this would be the case if you press the mic button and don't speak into the mic (no background noise) and no PL is used. The x-axis is frequency and the y-axis is amplitude. Since 146.130 MHz is the carrier or center frequency, the FM detector output is zero.

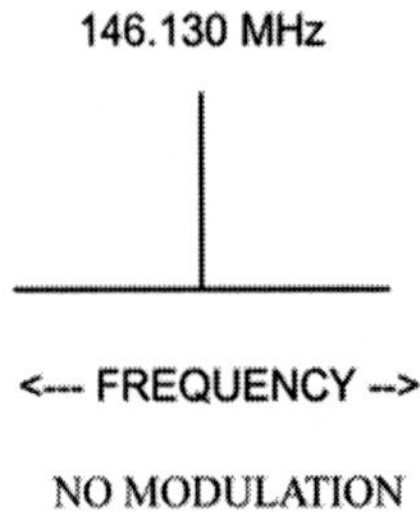

NO MODULATION

Assume that we modulate with a constant 1000-Hertz pure sine wave with the amplitude near the maximum allowed in repeater systems, then the RF signal would swing back and between 146.126 and 146.134 MHz at a 1000-Hertz rate. The carrier frequency increases during the positive cycle of the modulating signal and decreases during the negative portion. The difference between 146.126 and 146.134 MHz is +/- 4 kHz and is termed deviation. The modulating frequency could well be a PL tone e.g., 107.2 Hertz. but with less amplitude hence less deviation, typically +/- 500 Hz.

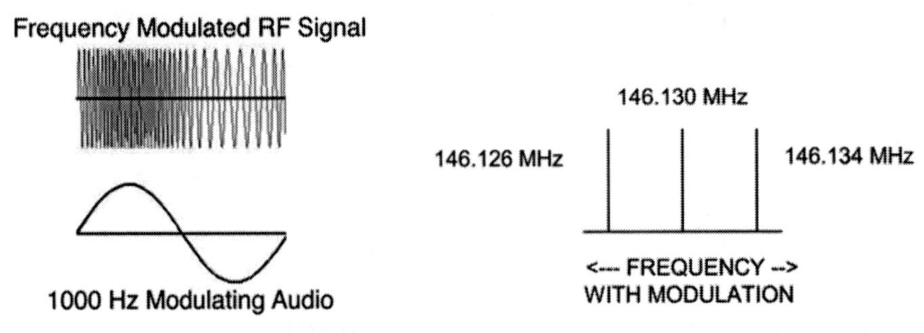

The illustration to the right is not what you would see on a Spectrum Analyzer, shown here only to illustrate deviation, not bandwidth, that will be covered later.

When the RF signal is passed through a discriminator it recovers the 1000 Hz audio and or the PL frequency.

This deviation is proportional to the AMPLITUDE of the modulating signal. Note that this does not affect the AMPLITUDE of the transmitted FM signal – only its frequency. When you speak softly the deviation is low, conversely shouting will increase deviation. If the radio is misadjusted or you shout too loud – the system deviation limits are exceeded and may result in distortion. In Amateur Radio FM radios, the maximum deviation is typically +/- 4.5 kHz. Deviation limits for CTCSS tones are typically +/- 500 Hz.

Over Deviation -- extremely loud audio with distortion on the audio peaks. Caused by improper internal adjustment and/or shouting into the microphone. Said radio or operator needs adjustment. Internally this is called a Dev adjustment. Not for the faint of heart on the micro-miniature component radios - but can be done. A great dynamic illustration of Frequency modulation and over deviation is at URL: http://www.williamson-labs.com/480_fm.htm

Under Deviation -- Symptomatic of soft-spoken persons. Everyone strains to hear them with volume controls wide open and mobiles with wind noise give up trying. When the next station comes in -- the wide open volume control blows listeners out of the room or car. Oouuch my ears!! Said radio needs a dev adjustment or a conscious effort of the operator to speak up or close-mouth the mic. Some folks cross talk the mic, speaking in to it at an angle rather than directly into it. This may take out some of the breathiness and make for a clearer audio.

Pre-emphasis of the audio frequencies is used in the transmitter (typically +6 dB/octave), this means that as the audio frequency doubles, the amplitude increases 6 dB. Conversely, to return the audio range to its normal response, de-emphasis of the audio frequencies are used in the receiver (typically -6 dB/octave) This is usually done between 300 - 3000 cycles.

Why is it necessary? Preemphasis is needed in FM to maintain good signal to noise ratio. Common voice characteristics emit low frequencies greater in amplitude than high frequencies. The circuits that clip the voice to allow protection of over deviation are usually not frequency sensitive, and are fixed in level, so they will clip or limit the lows before the highs. This results in added distortion because of the lows overdriving the limiter. Pre emphasis is used to shape the voice signals to create a more equal amplitude of lows and highs before their application to the limiter.

FM is not like AM or SSB, if two stations are transmitting at the same time (doubling) on the same frequency – it creates problems. If the two stations are about equal in strength to the repeater, both may be heard, typically alternately, with a heterodyne or beat note. When one of the stations is stronger than the other, the stronger station "captures" the weaker one and the strong guy wins. This is in contrast to AM or SSB where both may be heard. Expressed as capture ratio.

Another factor in FM modulation is the modulation index, which is the ratio of the frequency deviation of the modulated signal to the frequency of a sinusoidal modulating signal. Unlike AM modulation with two sidebands, FM can produce "extra" sidebands which are dependent on the modulation index. As the modulation index increases so does the number of sidebands. This gets into some complicated math using Bessel Functions. To reduce sidebands to an acceptable level and achieve reduced channel spacing, the modulation index is kept low (~1) in Amateur radio FM repeaters and is termed Narrow Band FM (NBFM).

This places restrictions on systems as far as bandwidth and frequency response are concerned. Bandwidth of an FM signal is given by 2 (delta F) + F_{Amax} where delta F is one half of the total frequency deviation and + F_{Amax} is the maximum audio frequency. Narrowband FM is described by the following general parameters: Note that repeaters usually have tighter specifications for channel spacing and guard channel requirements.

+/- 5.0 kHz peak deviation and a peak modulating audio frequency of 3.0 kHz.
The total bandwidth of the NBFM signal is calculated as:
Bandwidth (kHz) = (2 x Maximum Deviation) + (2 x Maximum Modulating Frequency)
or Bandwidth (kHz) = (2 x 5) + (2 x 3) or 16kHz

By contrast, FM broadcast stations (88 to 108 MHz) use wideband FM (WBFM), with a deviation of +/- 75 kHz, a bandwidth of 200 kHz, a modulation index ~5, and frequency response ~ 30 Hz to 15 kHz.

Many Ham systems use +/- 4 or 4.5 kHz peak deviation for narrower bandwidth. Using a +/- 4.0 kHz deviation gives a 15.5kHz total channel bandwidth. This has to do with allotted channel spacing, guard channel requirements, and compliance with local coordination specifications. Guard channels have been established which allow for transmitter drift and any small incidental sideband content.

After FM detection, a squelch circuit is used; this determines if there is sufficient audio and if not prevents any from reaching the transmitter. Some repeaters employ an audio gate – this prevents audio from modulating the transmitter if the audio has excessive noise and is weak. The incoming signal may be strong enough to hold up the repeater but no audio will get transmitted.

The controller is the brains of the system – it does everything from keying up and turning off the transmitter to configuring the autopatch, providing beeps, timers, and a whole lot more in between. Sophisticated controllers support voice mailboxes, weather reporting, signal strength and deviation reporting, multiple DTMF functions, alarms for loss of power, and smart autopatch handling. Auxiliary expansion ports can be implemented to connect multiple repeaters, links and remotes together.

Controllers allow for scheduling several events to occur at programmed times of the day. A scheduler can turn the repeater off and on, change the CTCSS, disable telephone usage or any user programmable feature at predetermined times during the day. A courtesy message is available to signal that the time clock is not set and the scheduler is disabled. Some controllers have digital potentiometers that can be remotely controlled. The transmitter audio level, telephone audio level and squelch can be remotely adjusted.

With this feature, trips to the repeater site to adjust the audio or squelch controls are reduced. Some controllers have SmartSquelch which knows the difference between a hand-held moving in and out of nulls and a fast mobile flutter and will not chop up the audio. SmartSquelch can also change the courtesy message when the signal becomes noisier. This is the weak signal courtesy feature.

The intent of this book is a very basic explanation of the repeater innards. There are lots of books for getting much deeper into repeater technical explanations. See ARRL http://www.arrl.org/catalog/
For more on FM theory – see URL: http://en.wikipedia.org/wiki/Frequency_modulation

The bottom part of Figure 2 shows the transmitter section. The power amplifier in a repeater is not as powerful as one might think, typically 10 or 20 Watts is sufficient for mountain top repeaters to cover a wide area.

Here are some typical 2-meter transmitter specifications:

Maximum Transmitter Frequency Error: +/- 300 Hz
CW/Digital Voice ID Deviation: +/- 2.5 kHz
CTCSS Deviation: +/- 500 Hz
Voice Deviation Only (NO CTCSS): +/- 4.0 kHz
Voice + CTCSS Deviation Combined: +/- 3.5 kHz And +/- 500 Hz = +/- 4.0 kHz
DTMF Deviation: +/- 3.0 kHz
Total Maximum NBFM TX Deviation: +/- 4.0 kHz

Here are some specs used in Southern California. Maximum peak deviation of the repeater transmitter should be no greater than 4.2 kHz and maximum modulating frequency of 3 kHz, with a 20 dB rolloff of the post-limiter filter at 4.4 kHz. Maximum transmitter frequency error: +/- 300 Hz. The -6 dB IF bandwidth should be no greater than +/- 6 kHz, and the -50 dB IF bandwidth should be no greater than +/- 10 kHz. Filters should be designed for 12.5 kHz channel spacing and should yield exceptional 15 kHz adjacent-channel rejection. Repeaters should use the minimum transmitter power necessary to reach the users. All new repeater coordination will require the use of some form of selective access (CTCSS or DCS).

Channel spacing on 2 Meters has changed over the years, with the older ones having spacing of 15 kHz. When the new sub band (144.50 - 145.50 MHz) was opened up in 1977, the channel spacing was set to 20 kHz for less interference between channels. Since so many of the repeaters on the "old band" would require major modification, the old channel spacing remains the same for the 146 - 148 MHz sub band. For channel spacing on other bands, see Chapter 4. These are typically 20 to 25 kHz spacing, depending on the band.

An invaluable use of a repeater is during emergencies. The ability to communicate over wide areas can facillate quick and accurate message handling. Should the commercial power fail, most repeaters have a battery back up system. Some repeaters will give an alert when on battery power – such as sending Morse "P". The battery sizes can be quite large and the battery chargers are very heavy duty. All this requires considerable maintence and clubs routinely have work parties to keep the machines operational.

Repeaters require constant maintenance and you might be invited to a repeater work party complete with pizza and "beverages". This is a great way to meet folks, tell Ham lies, and get to know the repeater workings. Hams either own or can borrow some very exotic test equipment from work. This is a nice introduction to test equipment such as oscilloscopes, spectrum analyzers, deviation meters, power and SWR meters, and lots more.

Being on mountaintops or buildings, repeaters are prone to lightning strikes and very elaborate grounding systems are used. Being in remote areas and subject to severe weather; wind, rain, snow, and ice can play havoc with an installation, so frequent maintenance is the order of the day. Best support a club with membership dues and assist when you can in repeater work parties.

CHAPTER 12: CROSS BAND REPEATING, LONG RANGE INTERTIE, VOIP (VOICE OVER IP)

CROSS BAND REPEATING

Cross band repeating (CBR) is a feature with some VHF-UHF dual band radios that simply repeats what it receives on one band and automatically retransmit it on the another band. This might be used when hiking or camping. Your vehicle is on a hill and you are in a valley below. You can take an HT with you and transceive to your mobile radio which will amplify and cross band repeat to a repeater or other users. Some HTs can control the mobile rig allowing frequency changes and voice feedback of the changes. Several configurations can be used; Simplex-To-Simplex CBR, Simplex-To-Repeater CBR, One-Way-CBR, and various modes of duplexing, simplex, half, full, etc. Be sure to read the FCC rules regarding cross band repeating. See Part 97.119 (a) Identification; Part 97.201 Auxiliary Stations.

There are just too many schemes and different radios that can do crossband to cover in detail here. Each manufacturer has their own method of implementing cross band repeat, so you will have to read your manual for how to set up and use the radio. Here are some very good websites: See AC6V website URL:http://ac6v.com/repeaters.htm#cross

For Auxiliary or Remote Base operation, the radio manufacturers have application notes for the various radios. See URL:http://ac6v.com/repeaters.htm#cross

LONG RANGE INTERTIE SYSTEMS

Repeaters can be linked so as to cover much wider areas than a single repeater. Coverage of several states is not unusual. Usually these require membership so as to cover the additional cost of operation. Some are amenable to casual use. Inquire when you first encounter them. For a typical intertie system, see the San Diego WIN System URL: http://www.winsystem.org/ Other examples can be found at URL: http://ac6v.com/sandiego.htm See the Cactus InterTie, CalZona Link Sys, Condor Connection, and the WALA Intertie Net.

VoIP (VOICE OVER IP)

Ham voice internet connections are a recent introduction and are great for everyone, but especially those that cannot get on HF because of license or antenna restrictions. They use a network protocol called VoIP (Voice over IP). These allow Amateurs to contact other Amateurs worldwide. Currently, modes are IRLP, ILINK, ECHOLINK, and WIRES II.

These allow world contacts via repeaters, your computer, and other means. There are over 800 IRLP repeaters in use and over 71,000 EchoLink users in 128 countries and these are growing rapidly. We will discuss the three most popular modes, IRLP, EchoLink and WIRES II.

INTERNET RADIO LINKING PROJECT (IRLP).

There are now nearly 800+ repeaters around the world connected by the internet through the Amateur radio internet radio linking project (IRLP) 24 hours per day, 7 days a week. The whole system is DTMF (touch-tone) controllable. The most remote repeater on the system is on Antarctica. To call a repeater on the system you dial a 4-digit DTMF code. Usage generally requires membership, but some are open to all Hams. The membership repeaters will usually host you to get the gist of IRLP usage. With IRLP, you talk through a local repeater, which sends the transmission along the internet to a distant repeater and you have Radio to Radio contact with stations across the world. Doesn't count for DXCC, but nevertheless, an exciting mode.

IRLP software takes the audio from a receiver and feeds it into a computer sound card. The computer converts the audio into ADPCM digital data (the same format used by the phone companies for Long Distance service). The Linux PC then converts this digital information into digital packets each assigned with IP addresses for the destination node. These packets now flow through the internet to the destination Linux PC where the packets are decoded then sent to the sound card and out to the transmitter microphone of the link radio which then transmits the audio out over the local repeater.

The transmitter is keyed as soon as these TCP/IP (Internet Protocol) packets start to arrive. As soon as the data stops the link radio automatically un-keys and the process reverses. An IRLP world wide node map system is shown at URL: http://www.irlp.net/

Using the IRLP Network

To connect to another node (repeater), dial a 4 digit access code. Within a few seconds the node will identify in voice with their callsign and location. If the node is currently connected, you will receive a message as to which connection the other node (repeater) is currently connected to. If you are near your computer, you can check the status of the desired node in real time by going to http://status.irlp.net This page updates every minute.

As with any linking system, IRLP is subject to some minor audio delays. These delays are caused by the amount of time it takes for the various radios to decode the Tone Squelch information. So best to slow down and be patient. When you have completed a QSO, announce your call and dial the OFF code. This can be a 4-digit code or on some machines 73 is used. A voice ID will indicate the link is disconnecting.

IRLP Reflectors

A reflector is a server that allows multiple nodes (repeaters) to be linked together at the same time. As of press time there were 5 reflectors located in Toronto, Saskatoon, Denver Colorado, Yellowknife and Sydney Australia. The most common world-wide reflector is in Denver CO.

Most reflectors are hosted by public service minded companies who have lots of bandwidth to support the requirement of a reflector.

The amount of bandwidth required for a reflector is directly related to the number of connected nodes. During one Sunday evening net, 25 nodes were connected.

ECHOLINK

Echolink is gaining popularity, to take a tour of this mode, see URL: http://www.echolink.org/el/
The program allows worldwide connections to be made between stations, from computer to station, or from computer to computer. There are more than 96,000 registered users in 128 countries worldwide!

EchoLink uses a computer interface to communicate with other amateur radio stations. EchoLink is easy to use and works with Microsoft Windows operating systems and can be used on virtually any PC with an Internet connection. Neither IRLP nor EchoLink need much bandwidth for good communication and both will work quite well with "dial-up" connections of 33k and above. (28k leaves holes and is frustrating.) And of course broadband is best.

EchoLink can be downloaded and installed in a matter of minutes and has been designed to permit a simple PC connection or the more elaborate connection which integrates an existing simplex or repeater station. Using the latter, one needs a RAS (Remote Access Service) configured for auto-dial or an "always" on Internet connection. There are repeaters, ((R)) and links ((L)) and just plain nodes. For nodes, no repeater or link is involved, hence no time out etc. Also the number of persons in a round table can be limited to a manageable number, like 3 or 4 or even just 2.

When you download and install the software, you will be asked for your amateur radio call and some additional information that serves to verify that the call entered in the EchoLink data base is "legitimate". Verified can take up to 24 hours. After your call is verified, you will be able to connect to the EchoLink servers and begin using a new and exciting form of communication using the Internet and your PC and/or radio. For further information – see URL: http://echolink.org/el/

If you want to connect your radio equipment to your computer, you will need a special interface. This allows the computer to control the PTT function of your transceiver, and (optionally) to accept and process DTMF commands from the receiver. EchoLink is designed to work with several different types of interfaces, including; WB2REM iLINK Board, VA3TO Enhanced iLINK Controller, RigBlaster, G3VFP Echolink Interface Controller. See URL: http://www.echolink.org/el/interfaces.htm You can check for current EchoLinkl logins at URL: http://www.echolink.org/el/logins.asp

Accessing EchoLink From Your Radio Through an Echolink Repeater.

1. First identify your station on the repeater and that you are attempting Echolink operation.
2. You can determine if the Echolink gateway is operational by keying a "*" DTMF tone. The gateway node will respond with status information.
3. Key the four or five digit station code for the station or repeater that you wish to link to.
4. After a few seconds delay, the Echolink node will respond with a "Connected" message if the link was successful. If the link was not successful, key a "#" to disconnect the gateway node.
5. Following a successful connection, announce your presence on the linked repeater by calling CQ.
6. During a QSO using linked repeaters, let the repeater tail completely drop before transmitting to avoid any timeout problems.
7. At the end of the QSO, disconnect the link by keying a "#" DTMF tone. The Echolink gateway will respond with confirmation that the link has been disconnected.
8. To test your audio, there is an echo conference server, called ECHOTEST. Once connected, the server records anything you transmit and plays it back. This is a convenient way to verify that your transmitted audio is clean, and to adjust record and playback sound levels. Use DTMF code list for the proper code.

Accessing Echolink From Your Computer

1. Download, install the iLink user program from the Echolink web site: http://www.echolink.org/
2. Connect a microphone to your computer and check for the proper microphone volume level.
3. Connect to the internet and run the Echolink user program.
4. Select a station or repeater to link to from the displayed Echolink station link.
5. Once connected, use the Enter key or spacebar to toggle between transmit and receive.

URL's for more Echolink information:

http://echolink.org/el/ Big Main Page For Echolink.
http://www.synergenics.com/sc/ EchoStation
http://groups.yahoo.com/group/echolink/ EchoLink Yahoo Discussion Group
http://www.ilinkboards.com/ VoIP Interface Boards
http://w0ant.s5.com/custom.html EchoLink Audio Adjustments
http://www.repeater.org/echolink/ EchoStation repeater control program for Windows which makes it easy to set up a complete, fully-functional repeater or "announcement machine" using a personal computer.

WIRES II (Wide-Coverage Internet Repeater Enhancement System)
From The Yaesu Site: HRI-100 URL: http://www.vxstd.com/en/wiresinfo-en/

WIRES-II uses DTMF signaling to establish a bridge, using the Internet, from your repeater or home station to another WIRES-II-equipped station anywhere in the world. At the repeater site, a personal computer is connected to the HRI-100 WIRES-II Interface Box, which serves as a command and audio-patching controller for the Internet bridge to your computer. Either a dial-up connection, or a high-speed line such as a DSL or ISDN line, may be used for connecting to the Internet. The flexibility of the WIRES-II concept allows you to configure the system to allow on-the-fly selection of linked or non-linked communications.

For fast-moving emergency communications where both local coordination and longer-distance reporting are required, WIRES-II allows local communications to be interspersed between linked transmissions. And because WIRES-II uses voice-recording technology as a buffer, WIRES-II calls will never interrupt a conversation in progress on a distant repeater. WIRES-II provides two operating network concepts.

Up to ten repeaters and/or home stations may join together to form a "Sister Radio Group" for closed-network operations, ideal for emergency, school, or sister-city groups. You can call any repeater within your SRG group using a single DTMF digit. And the host WIRES-II server also maintains a world-wide listing of repeaters operating in the "FRG" mode ("Friends' Radio Group"), any of whom you may call using a six-digit DTMF string to establish a link. More on WIRES II at URL:
http://k0swi.microlnk.com/IRADIO/WIRES/System.htm

D-STAR

Some ICOM radios are coming out with the D-STAR system, In D-Star, the air link portion of the protocol applies to signals traveling between radios or between a radio and a repeater. D-STAR radios can talk directly to each other without any intermediate equipment or through a repeater using D-STAR voice or data transceivers. The gateway portion of the protocol applies to the digital interface between D-STAR repeaters. D-STAR also specifies how a voice signal is converted to and from streams of digital data, a function called a codec. D-STAR codec is known as AMBE® (Advanced Multi-Band Excitation) and the voice signal is transmitted in the D-STAR system at 3600 bits/second (3.6 kbps). For more info on D-Star, see URL's:

http://www.icomamerica.com/amateur/dstar/dstar2.asp

http://www.arrl.org/news/stories/2005/12/14/1/?nc=1

And a D-Star System Map is shown at URL:
http://www.d-starradio.org/

NOTES

GLOSSARY AND INDEX

REPEATER TERMS, ABBREVIATIONS, AND JARGON

A (Alpha)

access code - A code to activate a repeater function e.g. auto patch, link etc. One or more numbers and/or symbols are keyed in with a telephone keypad and transmitted to the repeater. **Page 7-10.**

alligator - A repeater that transmits further than it can receive, big mouth, small ears! **Page 4-2.** Also applied to the repeater timer when it times out and cuts you off. "The Alligator got ya" **Page 4-6.**

AM - Amplitude Modulation

appliance operator: Hams who neither build nor experiment with radio equipment, but merely operate commercial equipment, perhaps without understanding how it all works.

APRS - Automatic Packet Position Reporting System

ARC – 1.) Amateur Radio Club. 2.) Military Designation for Avionics (Aviation Radio Composite)

ARES - Amateur Radio Emergency Service

ARRL - American Radio Relay League, the national amateur radio organization in the USA. This organization has represented Amateur Radio interests since the birth of ham radio. It is well worthwhile to consider membership for its technical and representational support to hams.

ASCII - American Standard Code for Information Interchange. The ASCII 7-bit code represents 128 characters including 32 control characters.

auroral propagation - Propagation above 30 MHz by means of refraction by highly ionized regions around the Earth's poles.

autopatch - (Repeater Term) a device that interfaces a repeater to the telephone system to permit repeater users to make telephone calls. Often just called a "patch." **Page 7-10.**

B (Bravo)

bandpass - range of frequencies permitted to pass through a filter or receiver circuit.

band-pass filter: circuit that passes a range of frequencies and attenuates signals above and below

base – 1.) a radio station located at a fixed location as opposed to a mobile or pedestrian. 2.) Used to identify the control location in a network of radio stations.

barefoot - transmitting with a transceiver alone and no linear amplifier base loading - A loading coil at the bottom of an antenna to achieve a lower resonant frequency.

beam - an antenna that gives a directional beam pattern. See Yagi. **Page 6-1.**

bleed over- Interference caused by a station operating on an adjacent channel

BNC - Coax connector commonly used with VHF/UHF equipment -- Bayonet Niell-Concelman (standard connector type used on COAX cable, named for its inventors).

boat anchor - antique ham equipment -- So named because of weight and size.

bootlegger - Someone, usually not a Ham but a wannabe, making up a callsign, one usually not in the callbook, and getting on the air. Sometimes it is someone who already bought a radio, took the test and flunked, and then gets on the air anyway**. Page 10-2**.

break - (Repeater Term) used to interrupt a conversation on a repeater to indicate that there is an urgent priority traffic. If non-urgent, simply interject your callsign. **Page 7-8.**

break break (Repeater Term) used to intercede in a conversation with an emergency. **Page 7-8.**

break break break (Repeater Term) – A Mayday Transmission, life and/or property at risk.

brick, a power amplifier used for VHF/UHF systems. **Page 6-4**.

broadcasting: transmissions intended for the general public. Broadcasting is prohibited on the Amateur Radio Bands, other than QSTs which of are of interest to all Amateur Stations, example W1AW code practice transmissions. Also a one-way transmission with no specific intended recipient.

C (Charlie)

call book - a publication or CD ROM that lists licensed amateur radio operators

calling frequency: A standard frequency where stations attempt to contact each other. Example -- 146.52 is the USA National FM simplex calling frequency. **Page 3-2.**

candy store -- ham term for the local Ham Radio Dealer.

capturing-- (Repeater Term) On a repeater if two stations transmit simultaneously, the signals mix in the repeater's receiver and results in a raspy signal. FM has a characteristic whereby the stronger signals "captures" and over-rides the weaker one. **Page 9-1.**

carrier - a pure continuous radio emission at a fixed frequency, without modulation and without interruption. Several types of modulation can be applied to the carrier, See AM and FM. **Page 11-3.**

cavity resonator - a tuned circuit using the physical resonance of one or more tuned cavities**. Page 11-1.**

CBA - Callbook Address

center frequency - The unmodulated carrier frequency of an FM transmitter. **Page 11-3.**

center loading - A loading coil at the center of an antenna to achieve a lower resonant frequency

channel –1.) (Repeater Term) the pair of frequencies (input and output) used by a repeater. 2.) Also refers to radio memories, as in memory channel. 3.) Can also be a simplex channel (one frequency).

channel spacing - the frequency spacing between adjacent frequency allocations - may be 50, 30, 25, 15 or 12.5kHz, depending upon the frequency and convention in use in the area of the repeater. **Page 4-8.**

clear -- used to indicate a station is done transmitting

closed repeater - a repeater whose access is limited to a select group (see open repeater). **Page 7-1**.

coax, coaxial cable a type of wire that consists of a center wire surrounded by insulation and then a grounded shield of braided wire. The shield minimizes electrical and radio frequency interference. 50-ohm and 72 ohm characteristic impedances are typical. **Page 2-5.**

co-channel interference - the interference resulting when a repeater receives signals from a distant repeater on the same frequency pair. **Page 7-1**

controller: (Repeater Term) the control system within a repeater -- usually includes turning the repeater on-off, timing transmissions, sending the identification signal, controlling the auto patch and CTCSS encoder/decoder functions**. Page 11-5.**

control operator - (Repeater Term) the Amateur Radio operator designated to "control" the operation of the repeater, as required by FCC regulations. **Page 11-2.**

Copy(ing) -- indication of how well communications are received. "I have a good copy on you" also used as a question, as in "did you copy" - understand all" copying -- used to indicate one is monitoring as in "I was copying the mail" which means I was listening in on the conversation.

COR (carrier-operated relay) - a device that causes the repeater to transmit in response to a received signal

courtesy beep - (Repeater Term) an audible indication that a repeater user may go ahead and transmit, usually resets the timer. **Page 4-5.**

coverage - (Repeater Term) the geographic area that the repeater provides communications. **Page 3-2**

CPS - Cycles Per Second, this terminology was replaced by "Hertz" (see "Hertz"). Also characters per second.

CQ - calling any amateur radio station, may be sent in CW, phone or some digital modes, not used on VHF/UHF FM Repeaters. **Page 7-2.**

cross-band: the process of transmitting on one band and receiving on another**. Page 12-1.**

CTCSS - (Repeater Term) abbreviation for continuous tone-controlled squelch system, subaudible tones that some repeaters require to gain access. Commonly called PL (A Motorola Trade Mark**). Page 4-4.**

CW - Continuous Wave, In truth a continuous wave is an unmodulated, uninterrupted RF wave. However in common usage refers to Morse code emissions or messages which is an interrupted wave.

D (Delta)

dB - Decibel (1/10 of a Bel); unit for the ratio of two power measurements. **Page 6-5.**

dBc - In terms of RF signals, dBc is Decibels relative to the carrier level.

dBd - Decibels above or below a dipole antenna. **Page 6-5.**

dBi - Decibels above or below an isotropic antenna. **Page 6-5.**

desense (desensitization): the reduction of receiver sensitivity due to overload from a nearby transmitter. **Page 9-3.**

deviation - The change in the carrier frequency of a FM transmitter produced by the modulating signal. **Page 11-3.**

deviation ratio - the ratio between the maximum change in RF-carrier frequency and the highest modulating frequency used in an FM transmitter. Also see modulation Index. **Page 11-3.**

digipeater (digital repeater) - a packet radio repeater

diplexer - A frequency splitting and isolation device. Typically used to couple two transceivers to a single or dual band antenna, thus allowing one to receive on one transceiver and transmit on the other transceiver. Typical application 2M and 440MHz transceivers into a dual band antenna for satellite work.

dipole - the basic antenna consisting of a length of wire or tubing, open and fed at the center. The entire antenna is ½ wavelength long at the desired operating frequency. This antenna is often used as a standard for calculating gain, dBd. **Page 6-2.**

director - an element in front of the driven element in a Yagi or Quad and some other directional antennas.

doubling -- (Repeater Term) On a repeater if two stations transmit simultaneously, the signals mix in the repeater's receiver and results in a raspy signal. FM has a characteristic whereby the stronger signals "captures" and over-rides the weaker one. **Page 9-1.**

driven element - antenna element that connects directly to the feed line.

dropping out - (Repeater Term) a repeater requires a minimum signal in order to transmit, when a signal does not have enough strength to keep the repeater transmitting, it "drops out". **Page 2-1.**

D-Star -- D-STAR radios can talk directly to each other without any intermediate equipment or through a repeater using D-STAR voice or data transceivers – **See Chapter 12**

DTMF - (Repeater Term) abbreviation for dual-tone multi-frequency, the series of tones generated from a keypad on a ham radio transceiver (or a regular telephone). Uses 2-of-7 or 2-of-8 tones; often referred to by Bell's trademark Touchtone. **Page 4-5.**

dual-band antenna - antenna designed for use on two different Amateur Radio bands.

dummy load - a device which substitutes for an antenna during tests on a transmitter. It converts radio energy to heat instead of radiating energy. Offers a match to the transmitter output impedance.

duplex - (Repeater Term) a communication mode in which a radio transmits on one frequency and receives on another (also see full duplex, half duplex, and simplex). **Page 4-3.**

duplexer - (Repeater Term) a device used in repeater systems which allows a single antenna to transmit and receive simultaneously. A very narrow band device which prevents the transmitter from overloading the receiver. **Page 11-1.**

DX - (noun) distant station; (verb) to contact a distant station. **Page 3-2.**

dynamic range: How well a receiver can handle strong signals without overloading; any measure of over 100 decibels is considered excellent.

E (Echo)

earth ground - a circuit connection to a ground rod driven into the earth

Echolink Uses a network protocol called VoIP (Voice over IP).This program allows worldwide connections to be made between stations, from computer to station, or from computer to computer. There are more than 96,000 registered users in 128 countries worldwide! See **Chapter 12.**

EIRP (effective radiated power referred to isotropic) - ERP plus 2.14 dB to correct for reference to isotropic.

elephant - a repeater that receives further than it can transmit, big ears, small mouth! **Page 4-2.**

elmer - a mentor; an experienced operator who tutors newer operators

eleven meters - currently the CB band, once a Ham band

ERP (effective radiated power) - radiated power, allowing for transmitter output power, line losses and antenna gain**. Page 6-2**.

eyeball - A face-to-face meeting between two ham radio operators.

F (Foxtrot)

FB - Fine Business, good, fine, OK

FCC Federal Communication Commission -- Administers and regulates the radio laws for the USA, sometimes called Uncle Charlie. **Page 1-2.**

field day - Amateur Radio activity in June to practice emergency communications. **Page 7-12.**

first personal - first name - CB jargon that has crept into Ham jargon - old timers shudder!

FM - Frequency Modulation. **Page 11-3.**

fox hunt - a contest to locate a hidden transmitter

frequency - the rate of oscillation (vibration). Audio and radio wave frequencies are measured in Hertz. (cycles per second)

frequency coordinator - (Repeater Term) an individual or group responsible for assigning frequencies to new repeaters without causing interference to existing repeaters. **Page 7-1.**

full duplex - a communications mode in which radios can transmit and receive at the same time by using two different frequencies (see "duplex" and half duplex). A telephone is full duplex. **Page 4-3**.

full quieting -- (Repeater Term) a phenomenon on FM transmissions where the incoming signal is sufficient to engage the receiver limiters - thus eliminating noise due to amplitude changes. **Page 11-3.**

G (Golf)

gain, antenna - an increase in the effective power radiated by an antenna in a certain desired direction, or an increase in received signal strength from a certain direction. This is at the expense of power radiated in, or signal strength received from, other directions. **Page 6-1.**

gateway - a link or bridge between one communication network and another. Can be repeater to satellite.

ground - Common zero-voltage reference point.

ground-plane antenna - a vertical antenna built with the central radiating element one-quarter-wavelength long and several radials extending horizontally from the base. The radials are slightly longer than one-quarter wave, and may droop toward the ground. **Page 6-1.**

ground wave propagation - radio waves that travel along the surface of the earth, even beyond the horizon. **Page 3-2.**

H (Hotel)

HAAT (height above average terrain) - a method of measuring antenna height.

half duplex - (Repeater Term) a communications mode in which a radio transmits and receives on two different frequencies but performs only one of these operations at any given time (see "duplex" and "full duplex"). **Page 4-3.**

half-wave dipole - the basic antenna consisting of a length of wire or tubing, open and fed at the center. The entire antenna is ½ wavelength long at the desired operating frequency. **Page 6-1.**

hand-held - (Repeater Term) a small, lightweight portable transceiver small enough to be carried easily; also-called HT (for Handie-Talkie, a Motorola trademark). **Page 2-1.**

Handi – Scratchy. Usually an HT transmission that barely accesses a repeater, resulting in heavy path noise and perhaps dropping in and out of the repeater.

hang time - (Repeater Term) the short period following a transmission that allows others who want to access the repeater a chance to do so; a courtesy beep sounds when the repeater is ready to accept another transmission**. Page 4-6.**

handle - A radio operator's name. Kind of unnecessary -- just say the name is. But lots of old timers use handle.

harmonic - a signal at a multiple of the fundamental frequency. Also a slang term for the children of an Amateur.

helical resonator - a compact resonant filter circuit to block multiple interfering signals.

HF - High Frequency - 3 MHz to 30 MHz

hi hi - ha ha (laughter) Originated in CW but often-used on voice.

homebrew - term for home-built, noncommercial radio equipment.

horizontally polarized wave - an electromagnetic wave with its electric lines of force parallel to the ground. In VHF/UHF - the antenna elements are horizontal (used at vhf/uhf for weak signal CW/SSB operation). **Page 6-1.**

HT - (Repeater Term) Handi-Talkie - a small hand held radio. **Page 2-1.**

I (India)

IC - Integrated circuit.

ID -- Identification, as announcing station callsign at intervals specified by Part 97 of the FCC Rules and Regulations**. Page 4-7.**

IF - Intermediate Frequency -- Intermediate frequency, resultant frequency from heterodyning the carrier frequency with an oscillator. Mixing incoming signals to an intermediate frequency enhances amplification, filtering and the processing of signals. Desirable to have more than one IF. **Page 11-3.**

image - A false signal produced in a superheterodyne receiver's circuitry. **Page 11-3.**

impedance: The opposition to the flow of electric current and radio energy; it is measured in ohms (symbol is Z). For best performance, the impedance of an antenna, the feedline, and the antenna connector on a radio should be approximately equal.

inductance - a measure of the ability of a coil to store energy in a magnetic field. Measured in Henrys.

inductor - an electrical component usually composed of a coil of wire wound on a central core. An inductor stores energy in a magnetic field.

input frequency - (Repeater Term) the frequency of the repeater's receiver (and your transceiver's transmitter). **Page 4-3.**

IRLP Radio Linking Project. Uses a network protocol called VoIP (Voice over IP).There are now nearly 800+ repeaters around the world connected by the internet through the Amateur radio internet radio linking project, 24 hours per day, 7 days a week. See **Chapter 12**

intermod - Short for "intermodulation," this means false or spurious signals produced by two or more signals mixing in a receiver or repeater station. Commonly used to describe the squeals and noise heard when using a high gain antenna on an HT. **Page 2-1.**

isolation – (repeater term) the difference in level (measured in dB) between a transmitted and received signal due to filtering.

isotropic - Theoretical "Single Point" antenna used to calculate gain. Isotrope - a theoretical antenna with zero dimensions and a spherical radiation pattern. Gain is -2.14 dB from dipole. **Page 6-2.**

J (Juliet)

J antenna (J pole) - a mechanically modified version of the zepp (zeppelin) antenna. It consists of a half-wavelength radiator fed by a quarter-wave matching stub. This antenna does not require the ground plane that ¼-wave antennas do to work properly. **Page 6-2.**

jam - cause intentional interference. **Page 10-1.**

jury rig -- Fix in an unorthodox manner

K (Kilo)

kc - (see "kilocycles")

kilo - the metric prefix for 10^3, or times 1,000

kerchunking - activating a repeater without identifying or modulating the carrier. **Page 9-3.**

key - (noun) any switch or button, usually refers to a telegraph or Morse code key

key - (verb) to press a key or button

key up - (Repeater Term) to turn on a repeater by transmitting on its input frequency.

key up - (verb) to activate a transmitter or repeater

L (Lima)

landline -- ham slang for telephone (lines).

LCD - Liquid Crystal Display

LED - Light-emitting diode

lid - a poor operator, one who does not follow proper procedures or sends sloppy Morse code.

limiter - (Repeater Term) a stage of an FM receiver that clips the tops of the FM signal thus makes the receiver less sensitive to amplitude variations and pulse noise. **Page 11-3.**

linear - an amplifier used after the transceiver output. So named for its purity of amplification. Linear, in the mathematical sense, means that what comes out is directly proportional to what goes in. As far as linear amps go, if you double the input, the output is doubled and so on. This does not generate any additional frequency byproducts. If the amp is nonlinear, sums, differences and all combinations of those are generated also. **Page 6-4**.

line-of-sight propagation - the term used to describe propagation in a straight line directly from one station to another. **Page 3-2.**

linking - the process of connecting repeaters in a permanent network, or one controlled by access codes.

M (Mike)

mA milliampere (1/1,000 ampere)

machine - a repeater

magnetic mount or mag-mount - (Repeater Term) an antenna with a magnetic base that permits quick installation and removal from a motor vehicle or other metal surface. **Page 2-4.**

mA/h - milliampere per hour

making the trip -- jargon for "successfully transmitting a readable message"

mega - the metric prefix for 10^6, or times 1,000,000.

megacycles - million cycles per second. This terminology has been replaced by MegaHertz (MHz)

megahertz - million hertz (see Hertz)

mic (mike) - microphone - a device that converts sound waves into electrical energy.

microwave - the region of the radio spectrum above 1 gigahertz (GHz).

milli - the metric prefix for 10^-3, or divide by 1,000.

mixer - a circuit that takes two or more input signals, and produces an output that includes the sum and difference of those signal frequencies. **Page 11-2.**

mW - milliwatt (1/1,000 watt)

mobile - an amateur radio station installed in a vehicle. **Page 2-4.**

mode – Usually indicates modulation type.

modem - short for modulator/demodulator. A modem modulates a radio signal to transmit data and demodulates a receive signal to recover transmitted data.

modulate - create a radio emission so that it contains information (voice, Morse code, music, binary, ASCII). Music is prohibited on Amateur Radio. **Page 11-3.**

modulation Index - (Repeater Term) the ratio between the maximum carrier frequency deviation and the audio modulating frequency at a given instant in an FM transmitter. **Page 11-4.**

motorboating -- an undesirable low frequency feedback resulting in a motorboat sound on the audio. **Page 9-2.**

multimode transceiver - transceiver capable of SSB, CW, AM, and FM operation. **Page 2-2.**

multipath -- reception over the direct path and a reflected path, causing reduced signals or distortion. **Page 3-2.**

mV - millivolt (1/1,000 volt)

N (November)

NB - Narrow band. Also noise Blanker

NBFM - narrow band FM. **Page 11-5.**

near field of an antenna - the region of the electromagnetic field immediately surrounding an antenna where the reactive field dominates and where the field strength as a function of angle (antenna pattern) depends upon the distance from the antenna. It is a region in which the electric and magnetic fields do not have a substantial plane-wave character, but vary considerably from point-to-point.

negative - no, incorrect

negative copy - unsuccessful transmission

negative offset - the repeater input frequency is lower than the output frequency. **Page 4-3.**

net - A group of stations that meet on a specified frequency at a certain time. The net is organized and directed by a net control station, who calls the net to order, recognizes stations entering and leaving the net, and authorizes stations to transmit. **Page 7-11.**

NiCad - Nickel Cadmium, generally refers to a type of rechargeable battery. **Page 6-6.**

NiMH - Nickel Metal Hydride, generally refers to a newer type of rechargeable battery. **Page 6-6.**

NTS - National Traffic System - an amateur radio relay system for passing messages. **Page 7-11.**

O (Oscar)

odd split - - (Repeater Term) unconventional frequency separation between input and output frequencies. **Page 4-3.**

offset - (Repeater Term) In order to listen and transmit at the same time, repeaters use two different frequencies. **Page 4-3.**

Ohm - The fundamental unit of resistance. One Ohm is the resistance offered when a potential of one Volt results in a current of one Ampere.

old man (OM) - friendship term, friend, pal or buddy. Sometimes "The Wise One"

open repeater - (Repeater Term) a repeater whose access is not limited. **Page 7-1.**

output frequency - (Repeater Term) the frequency of the repeater's transmitter (and your transceiver's receiver). **Page 4-3.**

over - used during a two-way communication under difficult copy - to alert the other station that you are returning the communication back to them. Not necessary on 2 Meter FM repeaters, as the courtesy beep serves this function. But using "over to Sam" is useful in a round table.

P (Papa)

packet radio - a system of digital communication whereby information is transmitted in short bursts. The bursts ("packets") also contain callsign, addressing and error detection information.

path noise - (Repeater Term) A term used in repeaters to indicate that the signal is so weak that the limiters have not engaged thus noise on the signal will be heard -- this is referred to as path noise and sometimes as (incorrectly) White Noise (See Same). **Page 9-2.**

PC - Printed Circuit

PEP - Peak envelope power - the average power of a signal at its largest amplitude peak.

personal - first name - as in "the personal here is Bob" -- CB jargon that has crept into Ham jargon - old timers shudder!

phase - the time interval between one event and another in a regularly recurring cycle.

phase modulation - varying the phase of an RF carrier in response to the instantaneous changes in an audio signal.

phone - voice modulation

phone patch - A connection between a two-way radio unit and the public telephone system. **Page 7-10.**

picket fencing - (Repeater Term) A condition experienced on VHF and above where a signal rapidly fluctuates in amplitude causing a sound akin to dragging a stick across a picket fence. If a repeater user's signal isn't strong enough to maintain solid access to the machine's input (such as when operating from a vehicle passing beneath underpasses or through hilly terrain), the signal would be hard to copy. **Page 9-1.**

pico - the metric prefix for 10^-12, or divide by 1,000,000,000,000.

pirate - Station using an existing callsign belonging to some else and operating illegally. **Page 10-2.**

PL - (Repeater Term) Private Line (same as CTCSS)-- low frequency audio tones used to alert or control receiving stations. PL, an acronym for Private Line, is Motorola's proprietary name for a communications industry signaling scheme call the Continuous Tone Coded Squelch System, or CTCSS. It is used to prevent a repeater from responding to unwanted signals or interference.

Tone is an electronic means of allowing a repeater to respond only to stations that encode or send the proper tone. Also used during the AutoPatch mode. **Page 4-4.**

Pled – refers to a device that employs Private Line (see above), typically a repeater that requires PL to access.

PLL- Phase-lock loop

PM - Phase Modulation, similar to Frequency Modulation

portable - A mobile is an amateur radio station installed in a vehicle - a mobile station can be used while in MOTION. A portable station is one that is designed to be easily moved from place to place but can only be used while stopped. Portable operation is away from the home base station.

positive offset - (Repeater Term) repeater input frequency is higher than the output frequency. **Page 4-3.**

pot - Potentiometer - Continuously variable resistor often used for adjusting levels, as in volume control.

PTT - Push To Talk, the switch in a transmitter circuit that activates the microphone and transmission circuitry

pull the plug -- shut down the station

Q (Quebec)

Q-Signals. Q-signals - a set of three-letter codes which are used by amateurs as abbreviations. Commonly used on both CW and phone. Note that many Hams use Q-Signals verbally, but they originated in CW communications, QTH is "my Location", "QSY is change frequency", etc -- See Q-Signals **Page 8-1.**

Q - A figure of merit for tuned circuits. For antennas, the Q is inversely proportional to useable bandwidth, with reasonable SWR.

quad - A directional antenna consisting of two one-wavelength "squares" of wire placed a quarter-wavelength apart.

QRP - Low power operation

QSL - to acknowledge receipt. Commonly used to indicate, "I understand", "I coped your transmission (or report) all OK". Also used as a term for sending cards by mail to confirm a two way contact with a station, such as QSL via the bureau.

QSO - Two-way conversation. **Page 8-1.**

Quagi -- an antenna constructed with both quad and yagi elements. **Page 6-3.**

R (Romeo)

RACES - Radio Amateur Civil Emergency Service. **Page 7-11.**

radio check - query from a station desiring a report on his stations signal strength and audio quality. **Page 7-9.**

ragchewing - chatting informally via radio

RDF - Radio Direction Finding

reading the mail - to listen to a QSO without participating

reflector - (1.an element behind the driven element in an Yagi and some other directional antennas. (2. IRLP Reflector -- a server that allows multiple nodes (repeaters) to be linked together at the same time. (3. a mail list that forwards news to the subscribers, e.g., DX Reflector, Contest Reflector.

repeater - A repeater is a receiver/transmitter that listens for your transmission and re-transmits it. Repeaters usually enjoy the advantage of height and power to extend the range of your transmission. Repeaters listen on one frequency and transmit on another. The separation between these two frequencies is referred to as the Offset. **Page 11-1.**

repeater directory - an annual ARRL publication that lists repeaters in the US, Canada and other areas. **Page 7-1.**

reset (repeater term) applies to when a repeater timer is reset back to zero and normally occurs when the carrier of the transmitter drops. **Page 4-6.**

reverse patch - when a call is received on its incoming telephone line this special autopatch rings over the air and may be answered by tone access.

rice box - equipment made in Japan or the Orient

rig - a radio (transmitter, receiver, or transceiver) Dates back to the 1950's

roger -- I understand - Received 100% In CW "R" dit dah dit

roger beep - a dit-dah-dit sent at the end of a transmission. **Page 4-6.**

rubber duck - A shortened flexible antenna used with hand-held scanners and transceivers. **Page 6-1.**

S (Sierra)

selectivity - Ability of a receiver to reject signals adjacent to tuned signal**. Page 11-3.**

sensitivity - A receiver's ability to receive weak signals. **Page 11-3.**

separation or split - the difference (in kHz) between a repeater's transmitter and receiver frequencies. Repeaters that use unusual separations, such as 1 MHz on 2 m, are sometimes said to have "oddball splits." **Page 4-3.**

shack - Ham station operating area

silent key - a deceased amateur operator

simplex - communications mode in which a radio transmits and receives on the same frequency. **Page 3-1.**

SK - Silent Key, an amateur term for indicating that a ham has passed away. Also one of the prosigns -- meaning "end of contact"

slim - Someone pretending to be a DX station, usually rare, that is supposed to be on the air. For example, someone in southern Argentina pretending to be on Heard Island. **Page 10-2.**

S-Meter - Signal Strength Meter

S/N - Signal-to-noise ratio

spectrum - the electromagnetic spectrum or some portion of it

split or separation - (Repeater Term) the difference (in kHz) between a repeater's transmitter and receiver frequencies. Repeaters that use unusual separations, such as 1 MHz on 2 m, are sometimes said to have "oddball splits" or odd split. **Page 4-3.**

spurs - Spurious Signals - Undesired signals and frequencies in the output of a transmitter.

SQL - Squelch - A circuit that mutes the receiver when no signal is present, thereby eliminating band noise. The squelch function is activated in the absence of a sufficiently strong desired input signal; in order to exclude undesired lower power input signals that may be present at or near the frequency of the desired signal.

squelch tail - (Repeater Term) A brief bit of noise heard between the end of a radio transmission and the reactivation of the receiver's squelch circuit.

SSB -- Single Sideband Suppressed Carrier. A voice mode where only one sideband either upper sideband (USB) or lower sideband (LSB) is transmitted.

sub-audible tone - see CTCSS. **Page 4-4.**

SWR - Standing Wave Ratio, a measure of how much radio energy sent into an antenna system is being reflected back to the transmitter. **Page 6-4.**

SWR meter - a device used to determine the Standing Wave Ratio of an antenna system. **Page 6-4.**

T (Tango)

tail - the brief signal transmitted by a repeater transmitter after someone stops talking.

third-party communications - messages passed from one amateur to another on behalf of a third person. **Page 1-2.**

TI - Talk-In Frequency

ticket - slang for an amateur radio license

timer - repeaters often incorporate a timer or transmit time limiter to control the length of a single transmission from a user. The time limit is set by the repeater owner. **Page 4-5.**

time-out - Excessively long transmission on a repeater causing the repeater's timer circuit to stop further transmissions (Slang - the alligator gottcha). **Page 4-2.**

tone pad -- an array of 12 or 16 numbered keys that generate the standard telephone dual tone multifrequency (DTMF) dialing signals. Like a standard telephone keypad. (see autopatch). **Page 7-10**

Touch-tone - trademark of AT&T. See DTMF.**Page 4-5.**

TNC - Threaded Niell-Concelman (standard connector type used on Coax cable, named for its inventors).

traffic - a message or messages sent by radio

transceiver - a radio that both transmits and receives

translator (linear translator) - a device used to directly convert and retransmit a block of received frequencies.

transponder - the term used for a linear translator in a satellite. Inverting transponder transmits received upper sideband as lower sideband. Non-inverting transponder transmits received upper sideband as upper sideband.

triggering - to activate a repeater by transmitting on its input frequency (see also key up).

triplexer -- see diplexer.

twisted pair - ham slang for telephone or telephone lines

U (Uniform)

UHF - Ultra High Frequency 300 - 3000 MHz

Uncle Charlie - The FCC. **Page 1-2.**

URL - Universal Resource Locator - Internet term

V (Victor)

VCO - Voltage-controller oscillator.**Page 4-4.**

VE - Volunteer Examiner, a person authorized to administer examinations for amateur radio licenses

VEC - Volunteer Examiner Coordinator, an amateur radio organization empowered by the FCC to recruit, organize, regulate and coordinate Volunteer Examiners.

vertical polarization - the antenna elements are vertical (For VHF/UHF used for FM and repeater operation). **Page 6-1.**

VFO - Variable Frequency Oscillator. In synthesized radios (most new ones) it is in discrete small steps rather than continuously variable. **Page 4-4.**

VHF - Very High Frequency 30 - 300 MHz

VSWR - Voltage standing wave ratio. **Page 6-4**

W (Whiskey)

white noise (Repeater Term) is a scientific term used to describe a spectrum of broadband noise generated in a receiver's detector and sampled to control the receiver's squelch. This term is often incorrectly used in repeater work to describe the sounds heard when the received transmission is noisy and hard to understand, usually attributed to a weak signal and the repeater limiters are not engaged. **Page 9-2.**

WIRES™-II (Wide-Coverage Internet Repeater Enhancement System) Uses a network protocol called VoIP (Voice over IP). It can link two repeaters or home stations, across a county or across a continent, thus allowing users with a hand-held or mobile transceiver to communicate over long distances. Nonproprietary DTMF signaling is used to establish the Internet link. **See Chapter 12**

wilco - Will comply

work - To communicate with another radio station, a valid two way contact

WPM - Words per minute; as in Morse code or typing speed

WX - weather

X (X-ray)

XCVR- Transceiver

XYL - Ex-Young Lady, wife

Y (Yankee)

Yagi - 1926 Hidetsugu Yagi and Shintaro Uda invented the "beam" antenna array. A directional antenna consisting of a dipole and two additional elements, a slightly longer reflector and a slightly shorter director. Electromagnetic coupling between the elements focuses maximum power (or reception) in the direction of the director. **Page 6-3.**

YL - Young Lady, any female amateur radio operator or the significant other of an amateur.

Z (Zulu)

zed - a phonetic for letter "Z"

NUMBERS 0-9

73 - Best regards;

88 - Love and kisses;

807 - Deceptive Ham term for a beer or liquid drink. Also a popular transmitting tube of the mid 1900's

FREE QUICK GUIDE TO FM REPEATER JARGON

Courtesy Of AC6V And N6FN

Good Golly Miss Molly, what is this all about?

One might hear "N4ZZZ this is K6XXX, Good morning OM, welcome to San Diego. Handle here is Jack, -- Juliet Alpha Charlie Kilo, QTH is El Cajon. You're not quite full quieting into the machine, about 20% path noise. Your deviation is fine. This repeater W6NWG Whiskey Six Nothing Works Good is located on Mount Palomar. The repeater gets very busy during commute hours so let's QSY to 146.075, plus offset with a PL of 107.2, QSL". This is followed by a beep, a quiet period, then the repeater drops off.

QRZ THE FREQUENCY – I'M QRU AND QRV – QSL - HUH

LID, DESENSE, CAPTURE, DEVIATION, BRICK, CTCSS, PATH NOISE

HUH

AC6V Publications
981 Texas Rd
Iola, KS
66749

REPEATER TERMS, ABBREVIATIONS, AND JARGON

A (Alpha)

access code - A code to activate a repeater function e.g. auto patch, link etc. One or more numbers and/or symbols are keyed in with a telephone keypad and transmitted to the repeater.

alligator - A repeater that transmits further than it can receive, big mouth, small ears! Also applied to the repeater timer when it times out and cuts you off. "The Alligator got ya"

appliance operator: Hams who neither build nor experiment with radio equipment, but merely operate commercial equipment, perhaps without understanding how it all works.

APRS - Automatic Packet Position Reporting System

ARES - Amateur Radio Emergency Service

ARRL - American Radio Relay League, the national amateur radio organization in the USA. This organization has represented Amateur Radio interests since the birth of ham radio. It is well worthwhile to consider membership for its technical and representational support to hams.

ASCII - American Standard Code for Information Interchange. The ASCII 7-bit code represents 128 characters including 32 control characters.

autopatch - (Repeater Term) a device that interfaces a repeater to the telephone system to permit repeater users to make telephone calls. Often just called a "patch."

B (Bravo)

bandpass - range of frequencies permitted to pass through a filter or receiver circuit.

band-pass filter: circuit that passes a range of frequencies and attenuates signals above and below

base – 1.) a radio station located at a fixed location as opposed to a mobile or pedestrian. 2.) Used to identify the control location in a network of radio stations.

barefoot - transmitting with a transceiver alone and no linear amplifier

bleed over- Interference caused by a station operating on an adjacent channel

BNC - Coax connector commonly used with VHF/UHF equipment -- Bayonet Niell-Concelman (standard connector type used on COAX cable, named for its inventors).

boat anchor - antique ham equipment -- So named because of weight and size.

bootlegger - Someone, usually not a Ham but a wannabe, making up a callsign, one usually not in the callbook, and getting on the air. Sometimes it is someone who already bought a radio, took the test and flunked, and then gets on the air anyway.

break - (Repeater Term) used to interrupt a conversation on a repeater to indicate that there is an urgent priority traffic. If non-urgent, simply interject your callsign.

break break (Repeater Term) used to intercede in a conversation with an emergency.

break break break (Repeater Term) – A Mayday Transmission, life and/or property at risk.

brick, a power amplifier used for VHF/UHF systems.

broadcasting: transmissions intended for the general public. Broadcasting is prohibited on the Amateur Radio Bands, other than QSTs which of are of interest to all Amateur Stations, example W1AW code practice transmissions. Also a one-way transmission with no specific intended recipient.

C (Charlie)

call book - a publication or CD ROM that lists licensed amateur radio operators

calling frequency: A standard frequency where stations attempt to contact each other. Example -- 146.52 is the USA National FM simplex calling frequency.

candy store -- ham term for the local Ham Radio Dealer.

capturing-- (Repeater Term) On a repeater if two stations transmit simultaneously, the signals mix in the repeater's receiver and results in a raspy signal. FM has a characteristic whereby the stronger signals "captures" and over-rides the weaker one.

carrier - a pure continuous radio emission at a fixed frequency, without modulation and without interruption. Several types of modulation can be applied to the carrier, See AM and FM.

center frequency - The unmodulated carrier frequency of an FM transmitter.

channel –1.) (Repeater Term) the pair of frequencies (input and output) used by a repeater. 2.) Also refers to radio memories, as in memory channel. 3.) Can also be a simplex channel (one frequency).

channel spacing - the frequency spacing between adjacent frequency allocations - may be 50, 30, 25, 15 or 12.5kHz, depending upon the frequency and convention in use in the area of the repeater.

clear -- used to indicate a station is done transmitting

closed repeater - a repeater whose access is limited to a select group (see open repeater).

coax, coaxial cable a type of wire that consists of a center wire surrounded by insulation and then a grounded shield of braided wire. The shield minimizes electrical and radio frequency interference. 50-ohm and 72 ohm characteristic impedances are typical.

co-channel interference - the interference resulting when a repeater receives signals from a distant repeater on the same frequency pair.

controller: (Repeater Term) the control system within a repeater -- usually includes turning the repeater on-off, timing transmissions, sending the identification signal, controlling the auto patch and CTCSS encoder/decoder functions.

control operator - (Repeater Term) the Amateur Radio operator designated to "control" the operation of the repeater, as required by FCC regulations.

Copy(ing) -- indication of how well communications are received. "I have a good copy on you"
also used as a question, as in "did you copy" - understand all" copying -- used to indicate one is monitoring as in "I was copying the mail" which means I was listening in on the conversation.

courtesy beep - (Repeater Term) an audible indication that a repeater user may go ahead and transmit, usually resets the timer.

coverage - (Repeater Term) the geographic area that the repeater provides communications.

CQ - calling any amateur radio station, may be sent in CW, phone or some digital modes, not used on VHF/UHF FM Repeaters.

cross-band: the process of transmitting on one band and receiving on another.

CTCSS - (Repeater Term) abbreviation for continuous tone-controlled squelch system, subaudible tones that some repeaters require to gain access. Commonly called PL (A Motorola Trade Mark).

CW - Continuous Wave, In truth a continuous wave is an unmodulated, uninterrupted RF wave. However in common usage refers to Morse code emissions or messages which is an interrupted wave.

D (Delta)

dB - Decibel (1/10 of a Bel); unit for the ratio of two power measurements.

desense (desensitization): the reduction of receiver sensitivity due to overload from a nearby transmitter.

deviation - The change in the carrier frequency of a FM transmitter produced by the modulating signal.

deviation ratio - the ratio between the maximum change in RF-carrier frequency and the highest modulating frequency used in an FM transmitter. Also see modulation Index.

diplexer - A frequency splitting and isolation device. Typically used to couple two transceivers to a single or dual band antenna, thus allowing one to receive on one transceiver and transmit on the other transceiver. Typical application 2M and 440MHz transceivers into a dual band antenna for satellite work.

doubling -- (Repeater Term) On a repeater if two stations transmit simultaneously, the signals mix in the repeater's receiver and results in a raspy signal. FM has a characteristic whereby the stronger signals "captures" and over-rides the weaker one.

dropping out - (Repeater Term) a repeater requires a minimum signal in order to transmit, when a signal does not have enough strength to keep the repeater transmitting, it "drops out".

DTMF - (Repeater Term) abbreviation for dual-tone multi-frequency, the series of tones generated from a keypad on a ham radio transceiver (or a regular telephone). Uses 2-of-7 or 2-of-8 tones; often referred to by Bell's trademark Touchtone.

dummy load - a device which substitutes for an antenna during tests on a transmitter. It converts radio energy to heat instead of radiating energy. Offers a match to the transmitter output impedance.

duplex - (Repeater Term) a communication mode in which a radio transmits on one frequency and receives on another (also see full duplex, half duplex, and simplex).

duplexer - (Repeater Term) a device used in repeater systems which allows a single antenna to transmit and receive simultaneously. A very narrow band device which prevents the transmitter from overloading the receiver.

DX - (noun) distant station; (verb) to contact a distant station.

dynamic range: How well a receiver can handle strong signals without overloading; any measure of over 100 decibels is considered excellent.

E (Echo)

earth ground - a circuit connection to a ground rod driven into the earth

Echolink Uses a network protocol called VoIP (Voice over IP).This program allows worldwide connections to be made between stations, from computer to station, or from computer to computer. There are more than 96,000 registered users in 128 countries worldwide! See

EIRP (effective radiated power referred to isotropic) - ERP plus 2.14 dB to correct for reference to isotropic.

elephant - a repeater that receives further than it can transmit, big ears, small mouth!

elmer - a mentor; an experienced operator who tutors newer operators eleven meters - currently the CB band, once a Ham band

ERP (effective radiated power) - radiated power, allowing for transmitter output power, line losses and antenna gain.

eyeball - A face-to-face meeting between two ham radio operators.

F (Foxtrot)

FB - Fine Business, good, fine, OK

FCC Federal Communication Commission -- Administers and regulates the radio laws for the USA, sometimes called Uncle Charlie.

field day - Amateur Radio activity in June to practice emergency communications.

first personal - first name - CB jargon that has crept into Ham jargon - old timers shudder!

FM - Frequency Modulation.

fox hunt - a contest to locate a hidden transmitter

frequency - the rate of oscillation (vibration). Audio and radio wave frequencies are measured in Hertz. (cycles per second)

frequency coordinator - (Repeater Term) an individual or group responsible for assigning frequencies to new repeaters without causing interference to existing repeaters.

full duplex - a communications mode in which radios can transmit and receive at the same time by using two different frequencies (see "duplex" and half duplex). A telephone is full duplex

full quieting -- (Repeater Term) a phenomenon on FM transmissions where the incoming signal is sufficient to engage the receiver limiters - thus eliminating noise due to amplitude changes.

G (Golf)

gain, antenna - an increase in the effective power radiated by an antenna in a certain desired direction, or an increase in received signal strength from a certain direction. This is at the expense of power radiated in, or signal strength received from, other directions.

gateway - a link or bridge between one communication network and another. Can be repeater to satellite.

ground - Common zero-voltage reference point.

ground-plane antenna - a vertical antenna built with the central radiating element one-quarter-wavelength long and several radials extending horizontally from the base. The radials are slightly longer than one-quarter wave, and may droop toward the ground.

ground wave propagation - radio waves that travel along the surface of the earth, even beyond the horizon.

H (Hotel)

half duplex - (Repeater Term) a communications mode in which a radio transmits and receives on two different frequencies but performs only one of these operations at any given time (see "duplex" and "full duplex").

hand-held - (Repeater Term) a small, lightweight portable transceiver small enough to be carried easily; also-called HT (for Handie-Talkie, a Motorola trademark).

Handi – Scratchy. Usually an HT transmission that barely accesses a repeater, resulting in heavy path noise and perhaps dropping in and out of the repeater.

hang time - (Repeater Term) the short period following a transmission that allows others who want to access the repeater a chance to do so; a courtesy beep sounds when the repeater is ready to accept another transmission.

handle - A radio operator's name. Kind of unnecessary -- just say the name is. But lots of old timers use handle.

harmonic - a signal at a multiple of the fundamental frequency. Also a slang term for the children of an Amateur.

HF - High Frequency - 3 MHz to 30 MHz

hi hi - ha ha (laughter) Originated in CW but often-used on voice.

homebrew - term for home-built, noncommercial radio equipment.

horizontally polarized wave - an electromagnetic wave with its electric lines of force parallel to the ground. In VHF/UHF - the antenna elements are horizontal (used at vhf/uhf for weak signal CW/SSB operation).

HT - (Repeater Term) Handi-Talkie - a small hand held radio.

I (India)

IC - Integrated circuit.

ID -- Identification, as announcing station callsign at intervals specified by Part 97 of the FCC Rules and Regulations.

IF - Intermediate Frequency -- Intermediate frequency, resultant frequency from heterodyning the carrier frequency with an oscillator. Mixing incoming signals to an intermediate frequency enhances amplification, filtering and the processing of signals. Desirable to have more than one IF.

image - A false signal produced in a superheterodyne receiver's circuitry.

impedance: The opposition to the flow of electric current and radio energy; it is measured in ohms (symbol is Z). For best performance, the impedance of an antenna, the feedline, and the antenna connector on a radio should be approximately equal.

inductance - a measure of the ability of a coil to store energy in a magnetic field. Measured in Henrys.

inductor - an electrical component usually composed of a coil of wire wound on a central core. An inductor stores energy in a magnetic field.

input frequency - (Repeater Term) the frequency of the repeater's receiver (and your transceiver's transmitter).

IRLP Radio Linking Project. Uses a network protocol called VoIP (Voice over IP). There are now nearly 800+ repeaters around the world connected by the internet through the Amateur radio internet radio linking project, 24 hours per day, 7 days a week.

intermod - Short for "intermodulation," this means false or spurious signals produced by two or more signals mixing in a receiver or repeater station. Commonly used to describe the squeals and noise heard when using a high gain antenna on an HT.

isolation – (repeater term) the difference in level (measured in dB) between a transmitted and received signal due to filtering.

isotropic - Theoretical "Single Point" antenna used to calculate gain. Isotrope - a theoretical antenna with zero dimensions and a spherical radiation pattern. Gain is -2.14 dB from dipole.

J (Juliet)

J antenna (J pole) - a mechanically modified version of the zepp (zeppelin) antenna. It consists of a half-wavelength radiator fed by a quarter-wave matching stub. This antenna does not require the ground plane that ¼-wave antennas do to work properly.

jam - cause intentional interference.

jury rig -- Fix in an unorthodox manner

K (Kilo)

kerchunking - activating a repeater without identifying or modulating the carrier.

key - (noun) any switch or button, usually refers to a telegraph or Morse code key

key - (verb) to press a key or button

key up - (Repeater Term) to turn on a repeater by transmitting on its input frequency.

key up - (verb) to activate a transmitter or repeater

L (Lima)

landline -- ham slang for telephone (lines).

LCD - Liquid Crystal Display

LED - Light-emitting diode

lid - a poor operator, one who does not follow proper procedures or sends sloppy Morse code.

limiter - (Repeater Term) a stage of an FM receiver that clips the tops of the FM signal thus makes the receiver less sensitive to amplitude variations and pulse noise.

linear - an amplifier used after the transceiver output. So named for its purity of amplification. Linear, in the mathematical sense, means that what comes out is directly proportional to what goes in. As far as linear amps go, if you double the input, the output is doubled and so on. This does not generate any additional frequency byproducts. If the amp is nonlinear, sums, differences and all combinations of those are generated also.

line-of-sight propagation - the term used to describe propagation in a straight line directly from one station to another.

linking - the process of connecting repeaters in a permanent network, or one controlled by access codes.

M (Mike)

machine - a repeater

magnetic mount or mag-mount - (Repeater Term) an antenna with a magnetic base that permits quick installation and removal from a motor vehicle or other metal surface.

mA/h - milliampere per hour

making the trip -- jargon for "successfully transmitting a readable message"

mic (mike) - microphone - a device that converts sound waves into electrical energy.

microwave - the region of the radio spectrum above 1 gigahertz (GHz).

mixer - a circuit that takes two or more input signals, and produces an output that includes the sum and difference of those signal frequencies.

mobile - an amateur radio station installed in a vehicle.

mode – Usually indicates modulation type.

modem - short for modulator/demodulator. A modem modulates a radio signal to transmit data and demodulates a receive signal to recover transmitted data.

modulate - create a radio emission so that it contains information (voice, Morse code, music, binary, ASCII). Music is prohibited on Amateur Radio.

modulation Index - (Repeater Term) the ratio between the maximum carrier frequency deviation and the audio modulating frequency at a given instant in an FM transmitter.

motorboating -- an undesirable low frequency feedback resulting in a motorboat sound on the audio.

multipath -- reception over the direct path and a reflected path, causing reduced signals or distortion.

mV - millivolt (1/1,000 volt)

N (November)

NB - Narrow band. Also noise Blanker

NBFM - narrow band FM.

negative - no, incorrect

negative copy - unsuccessful transmission

negative offset - the repeater input frequency is lower than the output frequency.

net - A group of stations that meet on a specified frequency at a certain time. The net is organized and directed by a net control station, who calls the net to order, recognizes stations entering and leaving the net, and authorizes stations to transmit.

NiCad - Nickel Cadmium, generally refers to a type of rechargeable battery.

NiMH - Nickel Metal Hydride, generally refers to a newer type of rechargeable battery.

NTS - National Traffic System - an amateur radio relay system for passing messages.

O (Oscar)

odd split - - (Repeater Term) unconventional frequency separation between input and output frequencies.

offset - (Repeater Term) In order to listen and transmit at the same time, repeaters use two different frequencies.

Ohm - The fundamental unit of resistance. One Ohm is the resistance offered when a potential of one Volt results in a current of one Ampere.

old man (OM) - friendship term, friend, pal or buddy. Sometimes "The Wise One"

open repeater - (Repeater Term) a repeater whose access is not limited.

output frequency - (Repeater Term) the frequency of the repeater's transmitter (and your transceiver's receiver).

over - used during a two-way communication under difficult copy - to alert the other station that you are returning the communication back to them. Not necessary on 2 Meter FM repeaters, as the courtesy beep serves this function. But using "over to Sam" is useful in a round table.

P (Papa)

packet radio - a system of digital communication whereby information is transmitted in short bursts. The bursts ("packets") also contain callsign, addressing and error detection information.

path noise - (Repeater Term) A term used in repeaters to indicate that the signal is so weak that the limiters have not engaged thus noise on the signal will be heard -- this is referred to as path noise and sometimes as (incorrectly) White Noise (See Same).

PC - Printed Circuit

PEP - Peak envelope power - the average power of a signal at its largest amplitude peak.

personal - first name - as in "the personal here is Bob" -- CB jargon that has crept into Ham jargon - old timers shudder!

phase - the time interval between one event and another in a regularly recurring cycle.

phase modulation - varying the phase of an RF carrier in response to the instantaneous changes in an audio signal.

phone - voice modulation

phone patch - A connection between a two-way radio unit and the public telephone system.

picket fencing - (Repeater Term) A condition experienced on VHF and above where a signal rapidly fluctuates in amplitude causing a sound akin to dragging a stick across a picket fence. If a repeater user's signal isn't strong enough to maintain solid access to the machine's input (such as when operating from a vehicle passing beneath underpasses or through hilly terrain), the signal would be hard to copy.

pirate - Station using an existing callsign belonging to some else and operating illegally

PL - (Repeater Term) Private Line (same as CTCSS)-- low frequency audio tones used to alert or control receiving stations. PL, an acronym for Private Line, is Motorola's proprietary name for a communications industry signaling scheme call the Continuous Tone Coded Squelch System, or CTCSS. It is used to prevent a repeater from responding to unwanted signals or interference.

Tone is an electronic means of allowing a repeater to respond only to stations that encode or send the proper tone. Also used during the AutoPatch mode.

Pled – refers to a device that employs Private Line (see above), typically a repeater that requires PL to access.

PLL- Phase-lock loop

PM - Phase Modulation, similar to Frequency Modulation

portable - A mobile is an amateur radio station installed in a vehicle - a mobile station can be used while in MOTION. A portable station is one that is designed to be easily moved from place to place but can only be used while stopped. Portable operation is away from the home base station.

positive offset - (Repeater Term) repeater input frequency is higher than the output frequency.

pot - Potentiometer - Continuously variable resistor often used for adjusting levels, as in volume control.

PTT - Push To Talk, the switch in a transmitter circuit that activates the microphone and transmission circuitry

pull the plug -- shut down the station

Q (Quebec)

Q-Signals. Q-signals - a set of three-letter codes which are used by amateurs as abbreviations. Commonly used on both CW and phone. Note that many Hams use Q-Signals verbally, but they originated in CW communications, QTH is "my Location", "QSY is change frequency", etc -- See Q-Signals

quad - A directional antenna consisting of two one-wavelength "squares" of wire placed a quarter-wavelength apart.

QRP - Low power operation

QSL - to acknowledge receipt. Commonly used to indicate, "I understand", "I coped your transmission (or report) all OK". Also used as a term for sending cards by mail to confirm a two way contact with a station, such as QSL via the bureau.

QSO - Two-way conversation.

Quagi -- an antenna constructed with both quad and yagi elements.

R (Romeo)

RACES - Radio Amateur Civil Emergency Service.

radio check - query from a station desiring a report on his stations signal strength and audio quality.

ragchewing - chatting informally via radio

RDF - Radio Direction Finding

reading the mail - to listen to a QSO without participating

repeater - A repeater is a receiver/transmitter that listens for your transmission and re-transmits it. Repeaters usually enjoy the advantage of height and power to extend the range of your transmission. Repeaters listen on one frequency and transmit on another. The separation between these two frequencies is referred to as the Offset.

repeater directory - an annual ARRL publication that lists repeaters in the US, Canada and other areas.

reset (repeater term) applies to when a repeater timer is reset back to zero and normally occurs when the carrier of the transmitter drops.

rice box - equipment made in Japan or the Orient

rig - a radio (transmitter, receiver, or transceiver) Dates back to the 1950's

roger -- I understand - Received 100% In CW "R" dit dah dit

roger beep - a dit-dah-dit sent at the end of a transmission.

rubber duck - A shortened flexible antenna used with hand-held scanners and transceivers.

S (Sierra)

selectivity - Ability of a receiver to reject signals adjacent to tuned signal.

sensitivity - A receiver's ability to receive weak signals.

separation or split - the difference (in kHz) between a repeater's transmitter and receiver frequencies. Repeaters that use unusual separations, such as 1 MHz on 2 m, are sometimes said to have "oddball splits."

shack - Ham station operating area

silent key - a deceased amateur operator

simplex - communications mode in which a radio transmits and receives on the same frequency.

SK - Silent Key, an amateur term for indicating that a ham has passed away. Also one of the prosigns -- meaning "end of contact"

slim - Someone pretending to be a DX station, usually rare, that is supposed to be on the air. For example, someone in southern Argentina pretending to be on Heard Island.

S-Meter - Signal Strength Meter

S/N - Signal-to-noise ratio

spectrum - the electromagnetic spectrum or some portion of it

split or separation - (Repeater Term) the difference (in kHz) between a repeater's transmitter and receiver frequencies. Repeaters that use unusual separations, such as 1 MHz on 2 m, are sometimes said to have "oddball splits" or odd split.

spurs - Spurious Signals - Undesired signals and frequencies in the output of a transmitter.

SQL - Squelch - A circuit that mutes the receiver when no signal is present, thereby eliminating band noise. The squelch function is activated in the absence of a sufficiently strong desired input signal; in order to exclude undesired lower power input signals that may be present at or near the frequency of the desired signal.

squelch tail - (Repeater Term) A brief bit of noise heard between the end of a radio transmission and the reactivation of the receiver's squelch circuit.

SSB -- Single Sideband Suppressed Carrier. A voice mode where only one sideband either upper sideband (USB) or lower sideband (LSB) is transmitted.

sub-audible tone - see CTCSS.

SWR - Standing Wave Ratio, a measure of how much radio energy sent into an antenna system is being reflected back to the transmitter.

SWR meter - a device used to determine the Standing Wave Ratio of an antenna system.

T (Tango)

tail - the brief signal transmitted by a repeater transmitter after someone stops talking.

third-party communications - messages passed from one amateur to another on behalf of a third person.

TI - Talk-In Frequency

ticket - slang for an amateur radio license

timer - repeaters often incorporate a timer or transmit time limiter to control the length of a single transmission from a user. The time limit is set by the repeater owner.

time-out - Excessively long transmission on a repeater causing the repeater's timer circuit to stop further transmissions (Slang - the alligator gottcha).

tone pad -- an array of 12 or 16 numbered keys that generate the standard telephone dual tone multifrequency (DTMF) dialing signals. Like a standard telephone keypad. (see autopatch).

Touch-tone - trademark of AT&T. See DTMF.

TNC - Threaded Niell-Concelman (standard connector type used on Coax cable, named for its inventors).

traffic - a message or messages sent by radio

transceiver - a radio that both transmits and receives

translator (linear translator) - a device used to directly convert and retransmit a block of received frequencies.

transponder - the term used for a linear translator in a satellite. Inverting transponder transmits received upper sideband as lower sideband. Non-inverting transponder transmits received upper sideband as upper sideband.

triggering - to activate a repeater by transmitting on its input frequency (see also key up).

triplexer -- see diplexer.

twisted pair - ham slang for telephone or telephone lines

U (Uniform)

UHF - Ultra High Frequency 300 - 3000 MHz

Uncle Charlie - The FCC.

URL - Universal Resource Locator - Internet term

V (Victor)

VCO - Voltage-controller oscillator.

VE - Volunteer Examiner, a person authorized to administer examinations for amateur radio licenses

VEC - Volunteer Examiner Coordinator, an amateur radio organization empowered by the FCC to recruit, organize, regulate and coordinate Volunteer Examiners.

VFO - Variable Frequency Oscillator. In synthesized radios (most new ones) it is in discrete small steps rather than continuously variable.

VHF - Very High Frequency 30 - 300 MHz

VSWR - Voltage standing wave ratio.

W (Whiskey)

white noise (Repeater Term) is a scientific term used to describe a spectrum of broadband noise generated in a receiver's detector and sampled to control the receiver's squelch. This term is often incorrectly used in repeater work to describe the sounds heard when the received transmission is noisy and hard to understand, usually attributed to a weak signal and the repeater limiters are not engaged.

wilco - Will comply

work - To communicate with another radio station, a valid two way contact

WPM - Words per minute; as in Morse code or typing speed

WX - weather

X (X-ray)

XCVR- Transceiver

XYL - Ex-Young Lady, wife

Y (Yankee

YL - Young Lady, any female amateur radio operator or the significant other of an amateur.

Z (Zulu)

zed - a phonetic for letter "Z"

NUMBERS 0-9

73 - Best regards; **88** - Love and kisses; **807** - Deceptive Ham term for a beer or liquid drink. Also a popular transmitting tube of the mid 1900's

Made in the USA
Charleston, SC
30 November 2009